D0638419

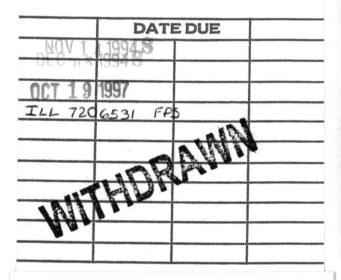

Geography of the Information Economy

GEOGRAPHY OF THE INFORMATION ECONOMY

MARK E. HEPWORTH

The Guilford Press
New York   London

© 1990 Mark E. Hepworth

Published in the USA and Canada by The Guilford Press
A Division of Guilford Publications, Inc.
72 Spring Street, New York, NY 10012

Printed in Great Britain

Last digit is print number: 9   8   7   6   5   4   3   2   1

Library of Congress Cataloging-in-Publication Data

Hepworth, Mark E.
    Geography of the information economy / by Mark E. Hepworth.
        p.   cm.
    Includes bibliographical references.
    ISBN 0–89862–786–9
    1. Information services industry—Location.   I. Title.
    HD9999.I492H46   1990
    338.6′042—dc20   89–23491
                                                                          CIP

Contents

List of Figures

List of Tables

Glossary of Technical Terms

Analogue/digital transmission
In analogue transmission, information is represented by a continuously variable signal; the signal is varied by discrete amounts in digital transmission.

Application/application program
A program that is designed to perform a particular application or set of tasks: e.g. word processing or data analysis.

Architecture (Network)
This defines the distribution of communications processing, data processing and data base management functions in a computer network; together with the hierarchical rules pertaining to message formats and link protocols.

Backbone network
The main system of nodes and links in a computer network. Terminals and access sub-networks are normally excluded.

Bandwidth
The frequency range of a communications channel.

Bit rate
The speed of data transmission in a telecommunications channel, say measured in bits per second.

Communications processing
The set of processing functions which ensure that data are exchanged between senders and receivers in the appropriate form, at the proper speed and without error.

Computer network
A generic term for a system of computers and terminals that can intercommunicate.

Concentrators
An intelligent device that enables data transmitted over a number of low-speed lines to converge and be immediately re-transmitted over a single high-speed line.

Data bank

An organised collection of information on 'files' which is computer-accessible.

Data communication

The transportation of encoded machine-readable information in a computer network.

Data processing

The computer-based processing of applications whereby value is added to data through arithmetical and logical operations.

Distributed processing

A network system in which information processing functions are geographically distributed between several 'hosts' at different locations.

Front end processor

A specialised computer connected to a mainframe computer which carries out communications processing functions on behalf of the latter.

Hierarchy (network)

A description of the multilevel structure of a computer according to how information processing functions are distributed.

Hosts (computer networks)

The primary computer/s in a network system on which application programs are run.

Inquiry/response systems

Transactions are transmitted frequently and response times are very short (a matter of seconds), e.g. credit authorisation.

Integrated Services Digital Network (ISDN)

A digital network which combines voice, data, facsimile and video in the same channels.

Interface (network)

The point of interconnection between systems and equipment in a computer network. Interfaces resolve any incompatibilities or provide a 'bridge' between the systems.

Leased (or dedicated) line

A telecommunications link for data or voice transmission that is reserved for the sole use of the leasing customer.

Modems

A device for converting digital signals (computer-generated) to analogue form, such that they are transmittable over an ordinary telephone line, and back again.

Multi-drop/point lines

A line which connects several terminals at different locations, thereby providing scale economies in data transmission through resource-sharing.

Multiplexers
A protocol-conversion device for combining and transmitting signals from a number of individual circuits over a common transmission path.

PAD machines
A device for interconnecting character-mode terminals to a packet-switched network.

Private/public network
Private networks are proprietary systems for authorised use by a single organisation or group of organisations; public networks operated by common telecommunications carriers are open to all subscribers.

Protocols
The rules governing communications in a computer network.

Remote access
The use of a computer from a terminal at a geographically distant point over telecommunications lines.

Remote job entry (or batch)
Transactions are usually submitted to a central computer in 'batches' and response-times in data processing may be a matter of hours or minutes, e.g. overnight transfer of daily sales records.

Switching
The means by which users are interconnected in a telecommunications network. Circuit-switched networks (e.g. the ordinary telephone system) provide a physical connection between subscribers for the entire duration of a call or session; message-switched networks employ store-and-forward techniques to route information between subscribers who time-share network links and modes; packet-switched networks are part of the last family except messages are fragmented into 'packets' consisting of a definite number of characters.

Topology (network)
The geographical configuration of a computer network which is represented for classification purposes by some geometric form, e.g. star or tree.

Value added network
A public network built with telecommunications links leased from the common carriers and offering enhanced services (e.g. network management).

Wide Area/Local Area Network
Wide area networks (WANS) are normally inter-city systems, whereas local area networks (LANS) are restricted to a specific site.

Editor's Preface

This is the first book in a new series. The series will attempt to map out the terrain of a new area of inquiry into what we have chosen to label as 'The Information Economy'. In this book Mark Hepworth sets the scene for the series by explaining what he means by the information economy and its particular geographical form. A subsequent volume will subject the term to further scrutiny from a variety of disciplinary perspectives. Other forthcoming contributions to the series will examine particular sectors of the information economy such as the media and cultural industries and research and development; the development of the infrastructure of the information economy in the form of advanced telecommunications; and major policy questions raised by the emergence of a single European 'Information Market' in 1992.

The series has its roots in a major research initiative of the Economic and Social Research Council (ESRC) in the UK which was launched in 1986 with the objective of examining the social and economic consequences of the widespread diffusion of information and communications technology. Six multidisciplinary research centres are engaged in this programme — at the Universities of Brunel, Edinburgh, Manchester, Newcastle and Sussex together with the Polytechnic of Central London. The Newcastle group is based in the Centre for Urban and Regional Development Studies (CURDS) and its work focuses on the implications of ICT for the development of cities and regions. Hence the subtitle of this series which will focus principally, although not exclusively, on research from CURDS and on area development issues.

The ESRC programme can in turn trace its origins to a growing concern amongst policy makers both within Britain and the OECD countries about the role of information activities in economic development. Within a wide range of policy arenas major decisions are having to be made in the absence of well grounded research on such issues as international trade in services (where the principal commodity is often information), on national and international investments in telecommunications infrastructure to transmit this information over space and on the regulation of information intensive activities like the media. Policy makers themselves are partly to blame for this lacunae having placed too much emphasis on the 'T' of 'IT' and having neglected the role of information generation, capture and transmission (or communication) in economic growth and change more generally. As Mark Hepworth clearly points out the demand for information and communications technology derives from a growing demand for information.

Not surprisingly there are exceptions to such generalisations about the orientation of public policy debate. For example the work of Porat for the OECD has attempted to determine the extent and significance of the information sector in member countries; his research provided the foundation for the burgeoning field of activity under the auspices of the Information, Computers and Communications Policy group at OECD which has sought to link information and technology policy questions to economic development. Within the UK the Government's Information Technology Advisory Panel, which, in a report entitled *Making a Business of Information*, analysed the contribution of tradeable information sector to the British economy and the role of ICT in supporting this sector is another exception.

A similar mis–orientation is apparent at the sub–national scale, particularly in relation to urban and regional development. A great deal of attention has been paid to the location of information technology production industries and to areas of high technology production like Silicon Valley in California and the M4 Corridor in England but far less work has been undertaken on the diffusion of ICT into a wide range of industries. While current debates do consider the greater flexibility that new technology is bringing to the organisation of production in different regions attention has tended to focus on the workplace and not the new flexible geography of organisations made possible by telecommunications. Partly as a result of the direction of research, policy initiatives have sought to create conditions for high technology production in lagging regions and ailing cities — for example through the promotion of science parks and its big brother 'the technopolis' and the creation of industrial districts based on a new flexible division of labour between firms.

An information economy perspective would suggest that this emphasis on research and policy fails to come to grips with the key dynamics in contemporary economic restructuring. It is as if the fundamental economic development issue in the 19th century had related to the location of steam engine production rather than the spread of steam power into a wide range of products and production processes (e.g. railway engines and steam driven looms) and the rapid improvements in communications that steam power made possible. Partly as a consequence of this lack of research on information – *and* communication technology many major policy decisions are being made in a range of fields with little regard for their spatial implications. And yet the spread of ICT in conjunction with the 'informatisation' of the economy carries with it implications for the location of activities and the development of cities and regions as profound as the spread of railways, roads and electric power. As it unfolds this series should shed some light on some of the uncertainties surrounding the geography of the future.

Before handing over the baton to Mark Hepworth it would seem appropriate for the Editor of the series to provide at least a preliminary definition of the Information Economy, as it relates to the development of places, particularly the city. As a starting point four interrelated propositions can be put forward. First, that information is coming to occupy centre stage as the key strategic resource on which the effective production and delivery of goods and services in all sectors of the world economy is dependent. Far from a transformation from an industrial to a post–industrial society where the emphasis is placed upon a shift from manufacturing to services, an Information Economy perspective would suggest that manufacturing and service activities are becoming equally dependent on effective information management. The city is — and always has been — the focus for information processing and exchange functions; as information becomes more important in both production and distribution, so the pivotal role of the city is reinforced.

The second proposition is that this economic transformation is being underpinned by a technical transformation in the way in which information can be processed and distributed. The key technical development is the convergence of the information processing capacity of computers (essentially a within workplace technology) with digital telecommunications (essentially a technology linking workplaces). The resultant technology of computer networking is emerging as a key spatial component in the technical infrastructure of the information economy. Because of their historic role cities are becoming the nodes or switching centres of this network based economy.

The third proposition is that the widespread use of information and communications technologies is facilitating the growth of the so called 'tradeable information sector' in the economy. This transformation embraces traditional information activities like the media, press, broadcasting and advertising, and the creation of new industries like on–line information services. Moreover many information activities previously undertaken within firms can now be purchased from external sources at lower cost in the 'information market place' — the growth of the advanced producer service sector can in part be accounted for by the externalisation of information functions from manufacturing and other firms. While the use of ICT permits an increasing volume of inter-organisation transactions, it seems that interpersonal contact remains sufficiently important, particularly in relation to the development of new services and relationships, that the role of cities is further enhanced.

The final proposition is that the growing 'informatisation' of the economy is making possible the global integration of national and regional economies. As the arena widens within which this highly competitive process of structural change is worked out, so the pattern of winners and losers amongst cities and regions is likely to become more sharply differentiated. Far from eliminating differences between places, the use of information and communications technology can permit the exploitation of differences between areas, for example in terms of local labour market conditions, the nature of cultural facilities and of institutional structures. It is therefore very important to see contemporary changes in a longer term historical perspective and in the context of the specificities of particular national space economies. Even in the information economy, geography matters!

John Goddard
Newcastle upon Tyne
July 1989

Acknowledgements

The origin of this book's main ideas is the PhD thesis I submitted to the University of Toronto in the summer of 1987. The thesis looked at selected aspects of the regional and international geography of the information economy in a Canadian context. I came to the topic via incidental studies of cybernetics in Allende's Chile, remote sensing technology and the 'international information order' and the geopolitics of information and transborder data flows. Later, by way of basic technical preparation for understanding IT, I studied for a year with Professor 'Kelly' Gotlieb of the Department of Computer Science in Toronto.

I am very grateful to Professor John Britton of the Department of Geography, University of Toronto, for his supervision of my PhD work, and also to Meric Gertler for his constructive comments on my ideas about 'information capital' and the thesis generally. Further thanks also extend to 'Kelly' Gotlieb, for his patient direction of my 'IT' studies and to Geoffrey Dobilas for his intellectual and personal companionship. Hoots laddie.

In September 1985, I joined the Centre for Urban and Regional Development Studies (CURDS), University of Newcastle upon Tyne, where geographical research on technological change has thrived for many years. I am particularly thankful to Professor John Goddard, the Director of CURDS, for giving me the dual opportunity of completing my PhD thesis and writing this book. These endeavours were helped considerably by CURDS' designation as a research centre in the Economic and Social Research Council's Programme on Information and Communication Technologies (PICT). I would like to thank the Centre's 'PICT Team' – most notably, Andrew Gillespie and Kevin Robins – for their critical contributions to my research pursuits. I also benefited greatly from my collaborations with Mike

Waterson (on the 'spatial dynamics of information technology': see Chapter Four) and Peter Monk (on 'conceptualising IT'), who have respectively moved on to the Universities of Reading and York.

I wish to thank Denise Weites for her excellent work in typing the original drafts of this book and other members of the CURDS secretarial staff for completing the job.

Finally, I owe a great personal debt to my wife, Jutta for her patience and morale boosting throughout the 1980s, and my parents for their devoted support of my academic work. Hello, Tanya.

August 1989   Mark Hepworth

Centre for Urban and Regional Development Studies (CURDS) University of Newcastle upon Tyne

This series is a joint venture between Belhaven Press and CURDS. The Centre was established in the University of Newcastle upon Tyne in 1977 to provide a focus for multidisciplinary research into urban and regional issues. CURDS became a Designated Research Centre of the Economic and Social Research Council (ESRC) in 1980 and in 1987 was one of the first of six centres to be funded under the Council's eight year Programme on Information and Communications Technologies (PICT).

PICT is the UK's first concentrated examination of the complex economic, social, managerial and policy issues raised by developments in information and communications technologies. In its contribution to PICT, CURDS is focusing on the implications of the communications revolution for the development of cities and regions in the UK through, for example, studies of the use of computer networks in organisations in different localities, particularly in relation to employment and trade in information based services, the geographical development of public telecommunications infrastructure and services and the information intensive media industries.

PICT is one of three major sub–groups in CURDS. The others are concerned with the development and application of the technology of Geographical Information Systems (funded under the auspices of the ESRC's Regional Research Laboratory initiative), and local economic and social development.

Professor John Goddard is Director of the Centre and there are currently 29 research staff and 7 support staff. Researchers have backgrounds in geography, economics, planning, management science, sociology, social anthropology, mathematics and computing. The Centre maintains a wide portfolio of externally financed research ranging from long term theoretical work through critical analysis of public policy to short term consultancy studies. It receives support from several directorates of the European Commission and the majority of UK Government departments. Its publications include: *The Northern Economic Review; Discussion Paper Series; Northern Regional Research Laboratory Reports; Newcastle Studies of the Information Economy.*

Chapter One
Introduction

Advanced industrialised economies are in the midst of an 'information revolution', whose geography will determine the future development prospects of the cities, regions and countries we live in. My book is about this geography. It looks at the spatial implications of recent changes in the economic locus of information, particularly as they relate to innovations in the converging technologies of computers and telecommunications. I have dubbed this broad field of research inquiry 'Geography of the Information Economy', and titled the book accordingly.

A contemporary study of the economic geography of information would, for obvious reasons, be of limited value if the role of information technology — broadly speaking, computers and telecommunications — were to be either down-played or even neglected. Indeed, a major part of this book is concerned with the geography of these technological innovations, whose profound significance for the present course of economic development is recognised both in academic circles and by policy-makers:

Information technology is all-pervasive. There is hardly a field now ... where [IT] does not in some way impact. It is widely held to be of crucial importance to development and economic success both in individual enterprises and in nations. The essential part it plays in commerce and industry, as well as in the daily lives of individuals, makes the policy governing it in the UK a matter of central importance. Japanese political and industrial leaders consider information technology to be the single most important factor involved in industrial and social development up to the end of this century and have given it the highest consideration in their national strategy. Aspects of information technology are major policy issues in the United States. Telecommunications and information technology have attracted the special attention of the European Commission in creating open markets'.[1]

Given that the demand for information technology (as new forms of capital) is basically a derivative relation, it is clear that we need to take account of the economic role of information itself. In this last respect, the 'newness' of information technology contrasts with the long economic history of informational resources, without which modern capitalist societies could simply not have evolved.

The information economy

The economic importance of information has grown steadily since the middle-nineteenth century period of industrial expansion in capitalist societies. This is attributed by Beniger (1986) to the rising 'economic demand for control', precipitated by 'increases in the speed of material processing and of flows through the material economy' (p. 435) generated by the Industrial Revolution. We are reminded by Beniger that, for example, the current electronic vintage of information-handling technologies has emerged from a long succession of what he calls 'control innovations' (Figure 1.1). At different historical

Year	Production	Distribution	Consumption	Generalised
1830		Scheduled		
32		freight line		
34			Penny press	
36				
38	Machine-tool	Telegraph		
1840	factory	Through-freight	Daguerreotype	
42	American System	forwarding	Advertising	Large-scale
44	of manufacture		agency	formal
46		Packaging	Hoe press	organisation
48	Standardised	Commodity	Newspaper	
1850	wire gauge	exchanges	association	
52	Commissioned	Postage stamp	Wood pulp, rag	Hierarchical
54	industrial	Through bill	paper	process con-
56	consultants	Registered mail	Iteration copy	trol system
58		Railroad scale	Typesetter	Formal line-
1860	Continuous-	Futures	Display type	and-staff
61	processing		Advertising	control
62	technology	Paper money	of Christmas	
63		Fixed prices		

Year	Production	Distribution	Consumption	Generalised
64		Postal money		
65		order	Premiums	
66		Transatlantic	for coupons	Modern bureau-
67		cable		cracies with
68	Bessemer	Travelling	Newspaper cir-	multiple de-
69	processing	salesmen	culation book	partments
1870	Continuous		Trademark law	
71	processing		Human-interest	
72	of materials	Mail-order	advertising	
73	Shop-order		Illustrated	Typewriter
74	accounting	Large chain	daily paper	with QWERTY
75	Plant design	of stores	Advertising	keyboard
76	to speed	Telephone	weekly	
77	processing			
78		Telephone	Full-page	
79		switchboard,	advertising	
1880		exchanges		
81			New trademark	Business school
82			law	Dow-Jones news
83		Uniform stan-	Mass daily	Accounting firm
84	Rate-fixing	dard time	Newspaper	Bonding company
85	department,	Special de-	syndicate	
86	cost control	livery mail	Linotype	Desktop
				telephone
87	Time recording			
88			Ad journal	Punch-card
89		Car accountant	National pub-	tabulator
1890	Staff time-	offices	licity stunt	Mimeograph
91	keepers for	Pay telephones,	Standardised	Multiplier
92	routing	travellers'	billboards	
93		checks	Print patent	Addressograph
94			Full-time	Four-function
95		Cafeteria	copywriters	calculator
96				Centralised,
97		Vending machine	Corporate pub-	departmental
98	Time studies	Rural free	licity bureau	corporate
99		delivery	Million-dollar	organisation
1900			ad campaign	Automatic card
01			Modern adver-	sorter
02		Automat	tising agency	Plug-board
03	Auto plant	Pacific cable	Advertising	tabulator
04	designed for		textbook	
05	processing	Wristwatches		
06	Factory control		Advertising	
07	by line-and-	Transatlantic	copy testing	
08	staff	radio		
09	Auto branch	Gyrocompass		
1910	assembly	Two-way auto	Formal market	Photostat
11	Scientific	radio	research	
12	management	Franchising	Mail-order and	

Year	Production	Distribution	Consumption	Generalised
13	Moving auto	Parcel post	testing	
14	assembly	Aircraft gyro-	Circulation	
15	Unattended	stabiliser	audit bureau	
16	substations	Self-service	Household	
17	River Rouge	store	market inter-	
18	processing	Air mail,	viewing	
19	architecture	Fedwire	Market	Printing
1920		Metered mail	research	tabulator
21		Drive-ins	textbook	Postage meter
22	Distant control	Shopping centre	Commercial	
23	of electrical	Supermarket	radio	Electric key-
24	transmission	Transcontinen-	National net-	punch
25	Demand feed-	tal air mail,	work radio	Decentralised
26	back control	facsimile	Dry waste	corporate
27	Pneumatic	Transatlantic	survey	organisation
28	proportional	telephone	Flasher sign	Multiple-
29	controller	Aircraft auto-	Radio ratings	register
		matic pilot		cumulating
				calculator
1930	Quality control		Automobile	
31	course, text	Teletype	radio	
32		service		
33	PID controller		Retail-sales	Machines linked
34	Pneumatic		index	for computing
35	transmitter		National polls	Electric
36	Lab analysis	Modern coaxial	Audimeter	typewriter
37	for quality	cable	ratings	
38	control	Radar	Animated sign	
39	Human relations	Transatlantic	Commercial	Electronic
	textbook	air mail	television	calculator

Figure 1.1 *Selected innovations in the control of production, distribution and consumption and in more generalised control, 1830–1939*

Source: J. Beniger, *The Control Revolution: Technological and Economic Origins of the Information Society*, 1986, pp. 430–1.

junctures, these innovations are argued to have played a crucial enabling role in capitalist economic development, through their delivery of the information requirements of industrial expansion based upon mass production, mass distribution and mass consumption.

As Figure 1.1 indicates, Beniger's list of 'control innovations' (up to 1939) includes not only new vintages of fixed capital but also new forms of industrial organisation. The latter, elaborated in great detail by Beniger elsewhere in his book, range from industrial specialisation (e.g. producer services, computer manufacture and telecommunications carriage), to the modern corporation (e.g. the 'U-form', then the 'M-form' hierarchy), to 'Scientific Management' on the shop-floor, and to a wide array of 'new' government regulatory agencies (e.g. standards bodies, marketing boards and statistics offices). What Beniger refers to as the 'Control Revolution' is, then, a historical succession of technological innovations ('old information technologies') *and* of extensions and refinements to the social and technical division of labour dedicated to the provision of informational resources at the microeconomic and macroeconomic levels (see also Jonscher, 1983).

The economic importance of these 'informational resources' was first assessed and highlighted by Machlup (1962), in his study of the US 'knowledge industry'. A major implication of Machlup's statistical evidence, relating to the 'GNP' and labour force shares of 'knowledge industries' and 'knowledge occupations', respectively, was that it invalidated one of the basic assumptions of traditional neoclassical theory — namely, that information can be treated as a costless resource (or, pure 'public good') in microeconomic and macroeconomic analysis. Thus, in a contemporary review of Machlup's work, Boulding (1963) remarked that 'the very concept of a knowledge industry contains enough dynamite to blast traditional economics into orbit' (p. 38). Some twenty years later, the effects of this 'blast' are still evident within the discipline:

At the time [1975], though economists had paid lip service to information, there was little formal literature. The last decade has seen burgeoning of the literature. It has become to the '70s and early '80s what growth theory was to the early '60s. And it has been treated with some of the same scepticism. [Stiglitz, 1985: 21]

Indeed, after its long occupancy of 'a slum dwelling in the town of economics' (Stigler. 1961: 61), information has become the central concern of an eclectic body of new economic theory. The following broad topics, according to McCall, Noll and Spence (1983), fall under the rubric of 'information economics':

The information sector of the economy — that is, a wide variety of specialised information industries and their regulation (computers, telecommunications, advertising, etc.).

The effects of differing amounts of information on the performance of product and factor markets (search theory, market signalling theory, etc.).

The implementation of macroeconomic and microeconomic policy under conditions of incomplete information or uncertainty (anti-trust 'transaction economics', monetary policy control, etc.).

The creation and use of new scientific and technological knowledge (R & D, patents, etc.).

In the course of this book, I will draw upon these various areas of study in order to develop some insights into the geography of the information economy. We should, however, note that interest amongst economists in the analytical importance of information has intensified, due mainly to a growing recognition of the fundamental weaknesses of traditional neoclassical theory. In this respect, the emergence of information technology has contributed to the exposure of these 'weaknesses':

Structural changes and the policy problems that emerge from them, the inherent weaknesses of theories that deny the analytical importance of the role of information, and the often spectacular events in the growth of "intelligent" electronics has generated much interest and, at long last, an economics of information. [Lamberton, 1983: 53]

The 'information economy' concept itself was coined by Porat (1977), although his definition and usage of the term requires some clarification. In closely following Machlup's sectoral approach. Porat assesses the relative size of the US 'information economy' in terms of the 'GNP' and labour force shares of 'information goods and services' and 'information occupations', respectively, based on inductive definitions of these analytical categories and a reclassification of statistical data from official national accounts and census sources. Purely on the basis of applying these measures of the 'information sector', Porat declares that, even by 1967, the USA could be described as an 'information economy':

We see that 46% of the Gross National Product is bound up with the information activity' [marketed and non-marketed information goods and services]; and [where] we discover that nearly half the labour force holds some sort of 'informational' job, earning 53% of labour income [1977:1]

However, what appears to contradict this declaration of findings is the following statement, which is to be found in the same text and very much as a confusing afterthought:

> By 1977, … we are entering another phase of economic history. We are just on the edge of becoming an information economy. The information technologies — computers and telecommunications — are the main engines of this transformation. And we are now seeing the growth of new information industries, products, services and occupations, which presage new work styles and lifestyles based on intensive use of information processing and communication technologies [Porat, 1977: 204]

My understanding of the information economy concept, as defined by Porat, is that it refers to a *new* phase of economic development, wherein the production of information goods and services dominates wealth and job creation with computers and telecommunications providing the technological potential for product and process innovation. As such, Porat's conception of the information economy fits neatly into the new 'techno-economic paradigm' adumbrated by Freeman (1987), Perez (1985), Miles (1988a) and other 'long-wave' theorists, falling within the time horizons of the so-called 'Information and Communication Kondratieff' (Figure 1.2). Also, as I have previously highlighted (Hepworth, 1986a), the information economy described by Porat is the economy of Bell's (1979) 'post-industrial society'.

The important aspect of Beniger's work is its historical perspective on the information economy, as a basic element of growth and specialisation in capitalist industrial development. This historical continuity is not captured by 'Kondratieff' analysis, or by the historical geography of information technology (Hall and Preston, 1988), or by Bell's 'post-industrial' schema. In essence, by making information rather than technological innovation the focus of his historical analysis, Beniger provides a more comprehensive and reliable account of the origins and secular development of the information economy, and its continuous relationship to so called 'control crises' in capitalist, mass-production economies.

By providing the information economy with a 'past', the 'shock of the new' is perhaps easier to absorb:

> Microprocessing and computing technology, contrary to currently fashion-

Approximate periodisation Upswing Downswing	Description	Main 'carrier branches' and induced growth sectors	Key factor industries offering abundant supply at descending price	Other sectors growing rapidly from small base	Limitations of previous techno-economic paradigm (Fordist mass production) and ways in which new paradigm offers some solutions	Organisation of firms and forms of cooperation and competition
1980s & 1990s to?	Information and communication Kondratieff	Computers Electronic capital goods Software Tele-communications equipment Optical fibres Robotics FMS Ceramics Data banks Information services	'Chips' (micro-electronics)	'Third generation' biotechnology products and processes Space activities Fine chemicals SDI	Diseconomies of scale and inflexibility of dedicated assembly-line and process plant partly overcome by flexible manufacturing systems, 'networking' and 'economies of scope'. Limitations of energy intensity and materials intensity partly overcome by electronic control systems and components. Limitations of hierarchical departmentalisation overcome by 'systemation' 'networking' and integration of design, production and marketing.	'Networks' of large and small firms based increasingly on computer networks and close co-operation in technology, quality control, training, investment planning and production planning ('just-in-time') etc. 'Keiretsu' and similar structures offering internal capital markets.

Figure 1.2 *The 'Fifth Information and Communication Kondratieff' (according to Freeman 1987)*

Source: C. Freeman (1987), *Technology Policy and Economic Performance*, London: Frances Pinter, Table 15, p. 75

able opinion, do not represent a new force only recently unleashed on an unprepared society but merely the most recent instalment in the continuing development of the Control Revolution. This explains why so many of the components of computer control have been anticipated, both by visionaries like Charles Babbage and by practical innovators like Daniel McCallum, since the first signs of a control crisis in the early nineteenth century [Beniger, 1986: 435]

In this book, I am exclusively concerned with *current* developments in industrialised countries that are associated with the growing economic importance of information, and, for this reason, the technological focus of attention is computer and advanced telecommunications innovations. What interests me, in particular, is what a spatial perspective might reveal about the various dimensions of the information economy, and what, in turn, this might imply for geographical research.

The scope of this book

The subject-matter of this book reflects, quite naturally, the specific research directions I have pursued until now. For example, though falling within the boundaries of the information economy, I have passed over the geography of several of its important constituent sectors — for example, mass media, education, 'large chunks' of the defence industries and information technology (IT) production itself. With respect to the last of the aforementioned, my work reflects an analytical bias towards the *use*, rather than the production of IT, and its spatial ramifications. Finally, to admit yet another caveat, this book provides a generalised view of the spatial aspects of the information economy, leaving aside its articulation with individual localities. I hope that, despite all these various shortcomings, my ideas on the information economy are of some value to other researchers.

We begin by establishing the fact that informational activities are of growing economic importance in industrialised countries (Chapter Two). I use Porat's 'occupational approach' for this purpose, given its advantage in revealing the rising information intensity of production (process-related changes) in all sectors, including manufacturing industry. The uneven spatial development of the information economy is highlighted in the specific cases of Canada and the UK. Finally, the ascendancy of 'information occupations' is contrasted with other significant trends in the labour market, particularly the parallel growth of lower order service jobs and so-called 'flexible work'.

In Chapter Three, attention shifts from the labour resources of the information economy to its capital resources, specifically information technology. It is shown here how the converging technologies of computers and telecommunications can be modelled as capital combinations with differing spatial structures. For this purpose, I invoke the concepts of information capital and computer network, and elaborate their spatial organisation. This descriptive analysis of information technology underlines the necessity of taking joint account of computer and telecommunications innovations in geographical research, but also indicates that 'mapping' information capital flows may prove to be extremely problematic. Attention is also drawn to the interrelated geographies of private and public computer networks, and their possible implications for differential access to the telecommunications infrastructure of the information economy. Lastly, it is argued that the informational economies of scale and scope associated with computer network innovations provide multilocational firms with enhanced production and locational flexibility.

The concept of 'informational capital' is elaborated further in Chapter Four, where I consider what its spatial attributes signify for conventional regional economic models — neo-classical production function, Keynesian income–expenditure and input–output approaches. Here, I emphasise the 'communicability' or real-time flow characteristics of information capital, and suggest that failure to spatially assign these capital service streams — mediated by computer networks — would severely undermine future attempts to model the urban and regional dynamics of the economy. In addition to offering further insights into the technological potential for 'flexible production', this chapter points to the new and different role played by telecommunications in regional economic development, as a primary channel for the spatial diffusion of (information) capital resources.

Chapter Five sees information capital 'in action' through some case study material on computer network innovations in large multilocational firms. Although modest in terms of their analytical depth and factual content, these studies bear testimony to the powerful capabilities of 'network' innovations for changing the geography of production, markets and work. The apparent locational flexibility of computer network mediated production does not, however, translate into a simple geography of 'centralisation versus decentralisation', given that either (or both) of these spatial tendencies arise according to the specific market and organisational contexts in which firms operate. This chapter, in addition, draws attention to the growing 'fuzziness' of the legal and administrative boundaries of the 'network firm' (Antonelli, 1988b), with the emergence of inter-organisational computer networks that link suppliers and purchasers — so-called 'electronic quasi-integration'.

Chapter Six pursues this last theme concerning the changing transactional structure of firms and its possible implications for the functional and spatial organisation of production. Here, I briefly review Coase's (1937) treatment of 'transaction costs' and theories of vertical (dis)integration, given their recent invocation by some economic geographers whose interests focus on 'flexible production' or 'flexible specialisation'. Here, I highlight the growing information intensity of the manufacturing firm and the analytical importance of information linkages in recent studies of the geography of 'flexible production'.

While the information economy's penetration of manufacturing production has attracted great interest amongst geographers, considerably less attention has been devoted to the dramatic 'collapse' of space and time which the use of information technology has precipitated in the world's capital markets. Chapter Seven offers my current, embryonic ideas on financial flexibility, which must surely be somehow introduced into the general debate on the international division of labour attendant to 'flexible production', rather than being 'excommunicated' to the producer services sector or the 'world city'.

In Chapter Eight, I have taken the opportunity afforded by writing this book to advance some further sketches of the 'information city' (see Hepworth, 1987c; Robins and Hepworth, 1988a). Here, I provide a critical interpretation of the 'electronic highways' (the advanced telecommunications infrastructure) and the 'wired city' concepts, whose advertisement by powerful commercial interests tends to obfuscate public debate on the political and social issues raised by urban development in the information economy. Also, drawing upon my current research pursuits, attention focuses on the information economy's expression within local government in the UK, where the privatisation of municipal services has highlighted both the economic and political importance of information in the present context of institutional reform.

The last chapter is a review essay, which attempts to draw together the numerous research themes presented in this book. What emerges from this reflective discussion are some of my own thoughts on the 'Geography of the Information Economy' as a developing field of research inquiry. In this regard, I view the book as a tentative and modest contribution to a complex business of agenda-setting, which is currently underway not only in geography but also in other social sciences.

Note

1. House of Commons, Trade and Industry Committee, *Information Technology* (2 Vols), London: HMSO, p. vi.

Chapter Two

Information in the Economy: An Occupational Perspective

Introduction

The historical origins of the information economy, as Beniger (1986) has demonstrated, can be traced back to the middle-nineteenth century period of industrial expansion. We have since witnessed a long succession of 'new' information technologies and organisational innovations, particularly the modern bureaucracy, induced by periodic 'crises of control' in capitalist economies founded on mass production, mass distribution and mass consumption systems. Indeed, in the current period, innovations in industrial organisation — so called 'flexible specialisation' (Piore and Sabel, 1984) — and information technology — broadly speaking, computers and telecommunications — are thought to offer solutions to the 'control problems' of a crisis-ridden, 'Fordist' mass production economy (Roobeek, 1987).

One important result of what Beniger calls the 'control revolution' has been a fundamental change in the occupational profile of the workforce in industrialised economies. This change in the structural division of labour, which parallels the historical development of new types of capital resources (technological innovations), has been marked by growth and specialisation in 'information occupations' — that is, occupations where job tasks primarily involve producing, processing and distributing information Machlup, 1962; Porat, 1977). The economic importance of these types of occupations in the United States (US) has grown over the last century and a half of industrialisation (Table 2.1); their post-war ascendancy in other advanced economies has been documented by the Organisation for Economic Co-operation and Development (OECD, 1981, 1986); the Canadian and British evidence on 'information occupation' trends presented in

Table 2.1 *Share of information occupations in the US labour force, 1800–1980**

OCCUPATIONAL SECTOR'S PERCENT OF TOTAL

Year	Agricultural	Industrial	Service	Information	Total labour force (in millions)
1800	87.2	1.4	11.3	0.2	1.5
1810	81.0	6.5	12.2	0.3	2.2
1820	73.0	16.0	10.7	0.4	3.0
1830	69.7	17.6	12.2	0.4	3.7
1840	58.8	24.4	12.7	4.1	5.2
1850	49.5	33.8	12.5	4.2	7.4
1860	40.6	37.0	16.6	5.8	8.3
1870	47.0	32.0	16.2	4.8	12.5
1880	43.7	25.2	24.6	6.5	17.4
1890	37.2	28.1	22.3	12.4	22.8
1900	35.3	26.8	25.1	12.8	29.2
1910	31.1	36.3	17.7	14.9	39.8
1920	32.5	32.0	17.8	17.7	45.3
1930	20.4	35.3	19.8	24.5	51.1
1940	15.4	37.2	22.5	24.9	53.6
1950	11.9	38.3	19.0	30.8	57.8
1960	6.0	34.8	17.2	42.0	67.8
1970	3.1	28.6	21.9	46.4	80.1
1980	2.1	22.5	28.8	46.6	95.8

Source: Beniger (1986), p. 24.
* Data are adapted from Porat (1977) and Bell (1979). See text for definition and measurement of the 'information sector' of the labour force and Appendix A for detailed taxonomy of occupations.

this chapter indicates that, at least for these countries, the 1970s and 1980s mark a further continuation of these secular developments in the labour force.

This chapter examines these occupational trends in some depth and suggests what they reveal about the current dimensions of structural change in advanced industrialised economies. It highlights the uneven development of the information economy in terms of the regional bias of growth and specialisation in 'information occupations' for the cases of Canada and the UK. The first sections offer a current definition of 'information occupations' and explain how they are accounted for as a basic indicator of structural change in the labour force.

Information occupations: definition and measurement

The terms 'information worker' and 'knowledge worker' were coined by Porat (1977) and Machlup (1962), respectively. There are, however, other partial equivalents of these occupational categories. Gottmann (1983), for example refers to 'quaternary' employment, defining quaternary services as 'closely related to the production, processing and distribution of information' (p. 66). Similarly, Gershuny and Miles (1983) point to the 'occupational tertiarisation' of manufacturing, or the growing information-processing component of the workforce within industrial sectors. At a much higher level of generality, information-handling occupations are (imprecisely) classified under the rubric of 'office', 'white-collar' or 'non-production' employment — with the last categorisation distinguishing productive and non-productive activities on the basis of their proximity (direct involvement) in the creation of physically tangible output.

These types of residual, broad or industry sector-based classifications do not meet our immediate needs of identifying information occupations in the labour force. They are motivated by, and reflect, different research concerns, which are mainly vertically (industry-sector) rather than horizontally (trans-sector) oriented. As Gershuny and Miles (1983), for example, are careful to point out:

> The new telecommunications and computing technologies of the 1970s are clearly having effects both on production processes and the development of new markets. These changes are not conveniently accommodated within the 'three sector' model. The new technologies draw our attention to a grouping of activities which do not fit comfortably into any of the traditional categories — activities concerned with information processing. Accordingly, some observers have added a fourth sector, the 'information sector', to the original three. [p. 248]

Let us now turn to this 'information sector', focusing on its occupational dimensions. We will concentrate on Porat's (1977) approach which, even with its various shortcomings (see Cooper, 1983: Hamelink, 1984), is the state of the art and most commonly applied version of 'information occupation' analysis.

The starting point of Porat's historical analysis of occupational trends in the US labour force is his operational definition of an 'information worker'. This is, he admits, a 'risky proposition' because all work requires some intellectual endeavour. Consequently, Porat's discussion of the concept is worth quoting at length:

> Does this worker's income originate primarily in the manipulation of symbols and information? Clearly, all human endeavour contains some component of information processing. Without information processing, all cognitive functions would cease and there would be no human activity. But, that definition is operationally useless. We are not saying that information workers deal exclusively in information and other kinds of workers never deal in information. Rather we assert that certain occupations are primarily engaged in the manipulation of symbols, either at a high intellectual content (such as the production of new knowledge) or at a more routine level (such as feeding computer cards into a card reader). And, for other occupations, such as in personal services or manufacturing, information handling appears only in an ancillary fashion. It is a distinction of degree not kind [pp. 105–6]

Based on this inductive definition, and the results of time-budget studies for numerous occupations,[1] Porat developed an inventory of 422 information occupations from the US Census of Population workforce classification. He then classified these occupations into four broad groupings — information producers, processors, distributors and infrastructure workers. The resulting taxonomy, as adapted for international comparisons by the OECD (1981, 1986), is presented in Table 2.2. (Appendix A contains a more detailed breakdown of these occupational listings). I have applied this same taxonomy, with minor adjustments for varying census job definitions, to identify information occupational patterns and trends for Canada (Hepworth, 1987a) and the UK. (Hepworth, Green and Gillespie, 1987).

The 'information sector' of the labour force, following Porat, is constructed by extracting information workers from all industries to form a single heterogeneous group. As such, the remaining components of the labour force are non-information workers in primary industries (e.g. miners and farm-hands), manufacturing and construction (e.g. welders and carpenters), and services (e.g. bus drivers and shop assistants). Several occupations are defined by Porat as 'ambiguous' — that is, they are not readily classifiable as information or non-information jobs. Examples of these '50:50' type of occupations include factory foremen and proprietors of small retail stores. All these ambiguous occupations are analysed here as 'informational', so that

Table 2.2 *Inventory of information occupations*

INFORMATION PRODUCERS	INFORMATION PROCESSORS	INFORMATION DISTRIBUTORS	INFORMATION INFRASTRUCTURE
SCIENTIFIC & TECHNICAL Examples: chemists & economists	ADMINISTRATIVE & MANAGERIAL Examples: production managers & senior government officials	EDUCATORS Examples: school & university teachers	INFORMATION MACHINE WORKERS Examples: computer operators & printing pressmen
MARKET SEARCH & CO-ORDINATION Examples: salesmen & buyers	PROCESS CONTROL & SUPERVISORY Examples: factory foremen & office supervisors	PUBLIC INFORMATION DISSEMINATORS Examples: librarians & archivists	POSTAL & TELECOMMUNICATIONS Examples: postmen & telegraph operators
INFORMATION GATHERERS Examples: surveyors & quality inspectors	CLERICAL & RELATED Examples: office clerks & bank tellers	COMMUNICATIONS WORKERS Examples: newspaper editors & TV directors	
CONSULTATIVE SERVICES Examples: accountants & lawyers			
HEALTH-RELATED CONSULTATIVE SERVICES Examples: doctors & veterinarians			

Notes: A complete list of information occupations is included in Appendix A. This is identical to the *OECD* inventory, except for the inclusion of public information disseminators and the separate classification of health-related consultative services. *Porat's* original inventory includes the same individual occupations, but the broad groups are organised differently: only clerical and related workers are treated as information processors; and, market search and coordination, information gatherers, administrative and managerial, and supervisory and control are all lumped together under one major group called 'Market Search and Coordination'. This study follows the OECD typology.

the estimates presented below should be treated as based on an 'inclusive' definition of the information labour force.

The 'information occupation' framework can be applied to individual industries, as a complementary perspective on changes in the character of production processes. It is, however, used by Porat to identify the industrial distribution of information workers at the conventional 'three-sector' level — primary, secondary and tertiary industries. For the Canadian case, I followed Bell's 'five-sector' model, which originates from Foote and Hart's (1953) refinements of Clark's (1940) path-breaking work (see Table 2.3). This level of sectoral aggregation is evinced somewhat by the following rationale provided by Beniger (1986):

Table 2.3 *A comparative schema of the post-industrial society (Daniel Bell)*

	Pre-Industrial	Industrial	Post-industrial
Mode of Production	Extractive	Fabrication	Processing, recycling
Economic Sector	Natural Resources	Manufacturing & Construction	Services:
			Tertiary:
			Transportation & Utilities
			Quaternary:
			Trade, Finance, Insurance & Real Estate
			Quinary:
			Health, Education, Research, Government, Recreation
Transforming Resource	Natural power	Created energy	Information Computer networks
Strategic Resource	Raw materials	Financial capital	Knowledge
Technology	Craft	Machine technology	Intellectual technology
Skill Base	Artisan, manual worker, farmer	Engineer, semi-skilled worker	Scientist, technical & professional occupation

Source: Bell (1979), pp. 166–7

Information processing and flows, increasing in response to the crisis in material control, needed themselves to be controlled, so that new control crises have appeared and been resolved throughout the twentieth century at levels of control increasingly removed from the processing of matter and energy.

The shop-order system of accounts based on routing slips developed in the mid-1870s to control material flows through factories, but by the 1890s a growing hierarchy of time-keepers and specialised clerks had become necessary to control not these throughputs themselves but their controlling flows of information. By the 1960s the hierarchies began themselves to fall under computer control requiring still newer information workers ... This (progressive) layering of control may also account for what Bell noted as the increasing importance of the quaternary and quinary sectors of advanced industrial economies: although the quaternary sector (finance, insurance and real estate) develops to control the extraction (primary sector), processing (secondary), and distribution (tertiary) of matter and energy, the quinary sector (law and government) develops to control the quaternary — that is, to *control* itself at a still higher level. [pp. 292–3]

Clearly, how the 'information occupation' framework is applied will depend on the research questions being immediately addressed. Indeed, later in this chapter, I will interrelate Porat's occupational analysis with the industry-sector taxonomy developed by Singlemann (1978), in order to reveal the informational base of the so-called 'new' service economy. What the framework does *not* imply, however, is that the direct employment impacts of the new information technologies will be restricted to information occupations. Indeed, what Porat classifies as 'non-information workers' — for example, shop assistants (point-of-sale terminals), truck drivers (cellular radio networks) and factory operatives (computer-controlled tools) — are also directly affected by the impact of the new technologies on work processes. The 'information occupation' approach should, therefore, be seen and used not as a technological forecasting tool, but as a conceptual framework for identifying basic changes in the transactional and productive structure of economies (or whole industries and firms).

Some evidence on national and regional trends

The main source of comparative evidence on national trends in information occupations is the OECD's statistical reviews (OECD,

1981, 1986). These 'information sector' accounts indicate that the share of information occupations in *all* OECD member countries has grown substantially in the post-war period (Table 2.4). There are, however, great differences in the sector's size between individual economies, varying from 23 per cent of the labour force in Norway to about 46 per cent in the US. Further, over the second half of the 1970s, the marginal (five-year) growth rate in information occupations declined for OECD countries as a whole (from 2.8 to 2.3 per cent).

The OECD evidence also indicates that, within the information labour force, there are significant inter-occupational differences in growth rates. The fastest growth rates are in management and administration, education and consultative services, but 'routine' information-handling occupations — mostly female-dominated clerical and related work — exhibited sharply declining growth rates through the 1970s. Finally, again notwithstanding international differences, about 70 per cent of the 1980–81 information labour force in OECD countries is concentrated in service industries, with manufacturing accounting for some 25 per cent on average. These sectoral and

Table 2.4 *Information occupations in the labour force of OECD countries: 1951–81*

	1951	1961	(per cent) 1971	1975	1981
Australia			39.4		41.5
Austria	18.0	22.0	28.5	32.2	
Canada	29.4	34.2	39.9		
Denmark					30.4
Finland	12.6	17.3	22.1	27.5	30.1
France	20.3	24.1	28.5	32.1	
Germany	18.3	23.4	29.3	32.8	33.5
Japan	17.9	22.2	25.4	29.6	
New Zealand				39.4	39.8
Norway				20.8	22.9
Sweden	26.0	28.7	32.6	34.9	36.1
United Kingdom	26.7	32.1	35.6		41.0
United States	30.7	34.7	41.1		45.8

Source: OECD, Update of Information Sector Statistics, Committee for Information, Computer and Communications Policy, February 1985.

occupational breakdowns for selected countries, as recorded by the
end of the last decade, are presented in Tables 2.5 and 2.6.
These post-war trends in the information labour force are, as
emphasised earlier, a continuation of occupational changes which date
back to industrialisation in the nineteenth century. In his unique

Table 2.5 *Occupational distribution of the information labour force in Canada
and other OECD countries: 1980–81*

OCCUPATION	CANADA	SWEDEN	(per cent) UNITED KINGDOM	UNITED STATES
Scientific/Technical	4.2	2.2	4.9	4.6
Consultative Services	12.5	3.9	8.0	8.5
Information Gatherers	3.0	0.5	3.2	0.9
Market Search	10.4	8.9	5.6	7.2
Administrative/Managerial	9.4	9.0	17.1	25.1
Process Control/Supervisory	13.3	20.8	8.8	8.7
Clerical/Related	33.6	29.6	32.9	31.9
Educators	9.0	13.6	9.5	9.2
Communication Workers	0.8	1.1	0.9	0.5
Information Machine	3.3	5.0	4.1	3.7
Postal/Telecommunications	2.0	7.2	4.9	3.0

Source: OECD (1985), Canadian figures from Hepworth (1987a)

Table 2.6 *Sectoral distribution of the information labour force in Canada and
other OECD countries: 1980–81*

SECTOR	CANADA[a]	SWEDEN	(per cent) GERMANY	UNITED KINGDOM[b]	FINLAND
Agriculture	1.1	1.1	0.3	0.3	1.7
Industry	21.4	28.2	29.6	30.6	26.2
Services	77.5	70.7	70.1	69.1	72.1

Source: OECD (1985); [a] Hepworth (1987a); [b] Hepworth, Green and Gillespie (1987).

historical account of the US information economy, Beniger (1986) traces the emergence of a significant information labour force to 'the crisis of control created by railroads and other steam-powered transportation in the 1940s', and 'the rapid bureaucratisation of the 1870s and 1880s' (pp. 23–4). Indeed, looking at the US data in Table 2.1, we can see that by 1940 over 50 per cent of the information sector's share of the American labour force was already established. This historical periodisation of occupational change, which led Beniger to conclude that the basic 'shape' of the information economy 'had been essentially completed by the late 1930s' (p. 293), contrasts with Bell's 'post-industrial' dating of the increasing centrality of information occupations. Stearns (1977), for example, underlines Beniger's view of the industrial origins of the information economy, as follows:

> Even in the information area (of the economy), the notion of elaborate personal records and a steady speed-up in communication was part of early industrialisation; therefore, one wonders whether we have not simply elaborated on this base.[p. 14]

This 'elaboration' of information-related activities is reflected in a highly articulated division of information labour within firms and government agencies — corporate and public bureaucracies — and between firms — market and quasi-market transactions in information-related services. Therefore, the growth of information occupations is spread across the entire economy, rather than simply being a service-sector phenomenon as Bell's 'post-industrial' schema of the 'information society' strongly suggests (see Table 2.3).

In Canada, for example, about 82 per cent of net growth in the national information labour force between 1971 and 1981 was concentrated in service industries (Hepworth, 1987a). Although secondary industries accounted for only 19 per cent of the total information workforce in 1981, their share was much greater for particular types of occupations — for example, Information Gatherers (46 per cent), Process Control and Supervisory (38 per cent) and Scientific and Technical occupations (26 per cent). Further, in terms of the occupational composition of their labour force, at least four major industries in the Canadian economy are classifiable as 'information-intensive' — that is, the workforce share of information occupations exceeds 50 per cent (for example, petroleum and coal products, chemical and chemical products and electrical products) (Table 2.7). What emerges from the Canadian evidence, at least, is a rising information intensiveness of production in all industries, including manufacturing.

This author's Canadian study also revealed a significant degree of *regional* variation in the size and growth rate of the information sector of the labour force at the level of Canada's ten provinces, nevertheless, the increasing dominance of information occupations was found to be a nationwide phenomenon. These common patterns of occupational change, emerging from highly specialised provincial economies in terms of their industrial base, are summarised in Table 2.8. Some disaggregated analysis of these regional trends revealed two basic aspects of Canada's 'information space' economy.

First, the growth pattern, size, and occupational composition of the information labour force in each Canadian province were strongly

Table 2.7 *Level and growth of information occupations in Canadian industries, 1971–81*

INDUSTRY	Information Workers in Total Labour Force (per cent)		
	1971	1981	Diff
1. Agriculture	6.0	13.4	7.4
2. Forestry	23.2	28.4	5.2
3. Fishing & trapping	3.9	8.6	4.7
4. Mines, quarries, oil wells	34.7	42.4	7.7
5. Food & beverage	31.1	30.6	–0.5
6. Tobacco products ·	38.2	46.7	8.5
7. Rubber & plastic products	38.9	38.1	–0.8
8. Leather	23.7	22.2	–1.5
9. Textile	31.9	33.1	1.2
10. Knitting Mills	28.6	29.1	0.5
11. Clothing	23.2	23.4	0.2
12. Wood	22.4	24.2	1.8
13. Furniture & fixtures	28.8	26.7	–2.1
14. Paper & allied	32.6	34.6	2.0
15. Printing, publishing & allied	77.1	90.5	13.4
16. Primary metal	31.3	33.1	1.8
17. Metal fabricating	35.6	35.3	–0.3
18. Machinery (excl. electrical)	50.6	49.3	–1.3
19. Transportation equip.	33.0	34.0	1.0

20. Electrical products	51.3	51.2	-0.1
21. Non-metallic mineral products	35.2	36.1	0.9
22. Petroleum & coal products	55.5	58.2	2.7
23. Chemical & chemical products	54.3	57.9	3.6
24. Misc. manufacturing	46.6	45.0	-1.6
25. Construction	23.6	30.2	6.6
26. Transportation, Communications, Utilities	50.2	53.8	3.6
27. Wholesale trade	61.7	62.3	0.6
28. Retail trade	42.8	45.5	2.7
29. Finance, Insurance, Real Estate	91.7	94.0	2.3
30. Education & related	84.5	86.0	1.5
31. Health & Welfare	32.3	35.4	3.1
32. Religious Orgs.	21.8	23.2	1.4
33. Amusement & recreation	39.8	44.8	5.0
34. Services to business management	85.8	86.0	0.2
35. Personal services	9.6	12.0	2.4
36. Accommodation & food	21.0	23.7	2.7
37. Misc. services	47.8	50.3	2.5
38. Public administration & defence	54.2	62.1	7.9

Source: Hepworth (1987a)

related to the industrial composition of the regional economy — manufacturing in Ontario/Quebec, extractive industries in British Columbia and the Prairie provinces (Alberta, Manitoba and Saskatchewan), and public services in the economically depressed Maritime provinces (Newfoundland, Prince Edward Island, Nova Scotia and New Brunswick). Second, and in addition to these occupational effects of regional industrial specialisation, the Canadian information economy is marked by a highly articulated 'spatial division of labour' (Massey, 1984) *within* individual industries. Although I have not attempted to investigate this second regionalisation factor directly, there is a wealth of facts on the spatially centralised structure of information-related activities in Canadian firms and in the country's federal government (see, for example: Coffey and Polese, 1987; Savoie, 1986).

The concept of a 'spatial division of information labour' has been invoked in other research carried out by Hepworth, Green and Gillespie (1987) on the regional distribution of information-related activities in the UK economy in 1981 (Table 2.9). It was shown here that the size and occupational composition of the information sector of

Table 2.8 *Net growth in the labour force by information and non-information occupations: Canada and Provinces, 1971–81*

CANADA AND PROVINCES	INFORMATION OCCUPATIONS (per cent) All Industries	NON-INFORMATION OCCUPATIONS (per cent) BY:				
		Primary Industries	Secondary Industries	Tertiary Industries	Quaternary Industries	Quinary Industries
CANADA	60.1	0.4	11.1	2.4	8.8	17.2
British Columbia	55.2	3.2	11.4	2.6	9.3	18.3
Alberta	59.7	2.0	12.4	3.8	8.6	13.5
Saskatchewan	67.0	–13.8	13.0	3.4	12.0	19.8
Manitoba	61.2	– 7.1	7.8	5.7	10.7	21.7
Ontario	61.7	0.3	10.5	2.8	8.1	16.6
Quebec	58.4	– 0.5	10.7	2.1	8.9	20.4
New Brunswick	52.0	3.1	14.4	1.0	9.5	20.0
Prince Edward Island	51.5	0.1	13.7	3.2	9.4	22.1
Nova Scotia	55.5	3.0	9.7	1.8	11.5	18.5
Newfoundland	44.5	6.9	19.4	0.5	9.4	19.3

Source: Hepworth (1987a)

Table 2.9 *The composition of the UK information workforce by region, 1981*

						Share of Total Labour Force (per cent)						
	UK	SE	East Anglia	Greater London	SW	West Midlands	East Midlands	Yorks & Humbs	NW	North	Wales	Scotland
Scientific & Technical	2.0	2.7	1.9	2.3	2.1	1.9	1.7	1.4	1.9	1.7	1.6	1.7
Market Search, etc.	2.9	3.2	2.5	4.2	2.7	2.9	2.5	2.5	2.6	1.9	1.8	2.2
Information Gatherers	1.7	1.6	1.5	1.6	1.5	2.3	2.0	1.7	1.8	1.6	1.5	1.6
Consultative Services	2.8	3.0	2.2	4.9	2.5	2.5	2.2	2.1	2.4	2.0	1.7	2.2
Health-Related Consultants	1.2	1.2	1.0	1.4	1.2	1.0	1.0	1.2	1.3	1.3	1.1	1.3
INFORMATION PRODUCERS	10.6	11.8	9.1	14.3	10.0	10.5	9.4	8.9	9.9	8.4	7.8	9.2
Admin. & Manag.: Govt.	0.9	0.9	0.9	1.5	1.1	0.5	0.6	0.7	0.7	0.8	1.0	0.9
Admin. & Manag.: Corporate	4.4	4.9	4.2	5.8	4.0	4.6	4.2	4.0	4.1	3.2	3.4	3.2
Managers & Proprietors	4.5	4.6	4.8	4.3	5.7	4.2	4.4	4.5	4.8	3.8	5.2	3.8
Process Control: Supervisory	1.4	1.6	1.3	1.9	1.3	1.1	1.1	1.2	1.3	1.4	1.4	1.5
Process Control: Foremen	2.4	2.1	2.5	1.7	2.2	2.5	3.7	2.8	2.6	3.0	2.8	2.5
Clerical & Related	13.9	14.5	12.4	19.5	13.1	12.5	11.5	11.9	13.5	12.2	11.5	12.6
INFORMATION PROCESSORS	27.6	28.5	26.1	34.7	27.4	25.5	25.5	25.1	27.0	24.4	25.3	24.5
Educators	3.4	3.6	3.2	2.9	3.3	3.5	3.3	3.6	3.6	3.5	3.7	3.5
Public Info. Distributors	0.1	0.1	0.1	0.2	0.1	0.1	0.1	0.1	0.1	0.1	0.1	0.1
Communication Workers	0.7	0.7	0.5	0.6	0.4	0.4	0.4	0.4	0.5	0.4	0.5	0.4
INFORMATION DISTRIBUTORS	4.2	4.5	3.8	4.8	3.9	3.9	3.7	4.1	4.2	4.0	4.3	4.1
Printing & Publishing	0.7	0.7	0.7	0.9	0.6	0.5	0.9	0.7	0.9	0.5	0.3	0.6
Equipment Operators, etc.	1.4	1.4	1.3	2.0	1.3	1.3	1.0	1.1	1.3	1.1	1.2	1.3
Postmen & Messengers	0.7	0.6	0.6	1.3	0.6	0.5	0.5	0.5	0.6	0.5	0.5	0.6
INFORMATION INFRASTRUCTURE	2.7	2.7	2.6	4.3	2.6	2.2	2.4	2.3	2.8	2.1	2.1	2.4
INFORMATION OCCUPATIONS	45.2	47.4	41.6	58.0	43.8	42.1	41.0	40.4	43.9	38.9	39.5	40.2

Source: Hepworth, Green and Gillespie (1987)

the labour force vary widely between the UK's standard planning regions (see also Hall, 1987). Further analysis indicated that this spatial variation can be attributed to the locational characteristics of information-intensive industries and inter-occupational differences in the spatial division of information labour across all industries. In particular, the national economy was shown to be characterised by the dominance of Greater London and the South-East region in knowledge production and head office management and control functions in the private and public sectors.

Johnson (1988) has extended this work on information occupational trends in the UK much further, using annual *employment* data for the 1975–84 period rather than decennial census data which describe a 'fully employed' workforce. Over this recent period, when registered unemployment in the UK doubled to reach a figure of over 3 million people, Johnson found that the share of the national employed labour force in information occupations increased from about 40 to 47 per cent (Table 2.10). These upward trends in the information sector of the labour force apply not only to the relatively prosperous, service and 'high-tech' economy of the South of England, but also to the economically depressed, manufacturing regions of the UK's so-called 'Periphery' and 'Heartland'.

Further analysis of specific information occupational groups and their inter-regional distribution revealed the following tendencies:

Table 2.10 *Share of information occupations in regional employment: UK, 1975–84*

Region	1975	1977	1979	1981	1983	1984
South[a]	43.1	43.0	44.2	49.7	49.5	52.4
Heartland[b]	41.8	43.1	43.9	42.5	42.2	44.9
Periphery[c]	34.4	35.3	38.1	39.3	40.0	41.6
UK	40.1	41.1	42.6	44.7	44.7	47.1

Notes:
[a] Includes the South-East of England, South-West and East Anglia.
[b] Includes the Midlands, North-West, Yorkshire and Humberside.
[c] Includes the North, Scotland, Wales and Northern Ireland.

Source: Johnson (1988)

Table 2.11 Employment change in the UK workforce, 1979–84: by region and information/non-information occupations

	1979		1984		Percentage Change 1979-84
	Thousands	Per cent of Workforce	Thousands	Per cent of Workforce	
*SOUTH**					
Information Workers	4636	42.6	5599	48.0	+20.8
Non-Information Workers	5852	53.8	5087	43.6	-13.1
Unemployed	386	3.5	969	8.3	+151.0
Total Workforce	10874	100.0	11655	100.0	+7.2
*HEARTLAND**					
Information Workers	3730	41.4	3743	39.1	+0.3
Non-Information Workers	4767	52.9	4593	48.0	-9.0
Unemployed	515	5.7	1226	12.8	+138.1
Total Workforce	9012	100.0	9562	100.0	+6.1
*PERIPHERY**					
Information Workers	1800	35.2	1800	35.7	0
Non-Information Workers	2925	57.2	2528	50.1	-13.6
Unemployed	390	7.6	716	14.2	+83.6
Total Workforce	5115	100.0	5044	100.0	-1.4
UNITED KINGDOM					
Information Workers	9886	39.5	11142	42.4	+12.7
Non-Information Workers	13824	55.3	12208	46.5	-11.7
Unemployed	1291	5.2	2911	11.1	+125.5
Total Workforce	25001	100.0	26261	100.0	+5.0

*See Table 2.10 for standard regions included.

Source: Johnson (1988)

significant growth in scientific and technical, managerial and consultative services workers; even more rapid growth in supervisory and foreman occupations related to process control in offices, factories and other workplaces; and, a declining share of the labour force in routine information processing occupations, particularly in the 'clerical and related' category. In light of observed regional differentiation in these occupational changes, Johnson points to the intensifying 'Southern' bias of the spatial division of information labour in the UK, which emerges most clearly from his focused analysis of the 1979–84 period — the first five years of Conservative administration — (Table 2.11).

In sum, the available evidence on national and regional trends points to the seemingly 'universal' ascendancy of information occupations in the labour force of advanced industrialised countries. We have, however, seen that the present size of the information sector (by occupation) varies considerably between and within countries, and, at least in the British and Canadian cases, the inter-regional division of information labour favours the established 'centres' of these economies. Finally, as Beniger's (1986) study of the US economy makes clear, our understanding of the growth of information occupations would benefit from historical research on particular countries and regions, given the unique position and experience of different places in the national and international division of labour.

Complementary perspectives

The 'information economy' model formulated by Jonscher (1983) attributes the secular growth of information occupations to differential rates of technological progress in the production and information sectors. According to Jonscher:

> As industrial technology develops, the processes of production leading to the final output of goods and services in the economy becomes more complex. The organisational or informational task of coordinating the diverse steps in the production chain grows, as the number of *transactions* within and among productive units increases. Since the functions of information handling have not benefited from comparable efficiency improvements, the number of information workers must grow in response to this increasing organisation task. [p. 15]

The 'organisation task' of controlling the mass production economy has, in addition to a large array of technical innovations, been handled by private and public bureaucracies. Both Porat (1977) and Beniger (1986) view these bureaucracies as the most important control technology of advanced industrial economies, and as the dominant institutional form taken by the 'division of information labour'. After Arrow (1979), Alchian and Demsetz (1972), and Marschak (1968), for example, Porat emphasises the role of vertical integration in explaining the growth of bureaucratic organisation: 'firms integrate vertically to economise on the information flows necessary to coordinate complex productive activities' (p. 151).

In recent years, the role of 'transaction costs' in explaining the structure of market and non-market forms of economic organisation has been subject to increasing attention both by economists (e.g. Williamson, 1975, 1985) and geographers (e.g. Scott, 1983; Storper and Christopherson, 1987). The informational basis of transaction costs, as Dahlman (1979) points out, is revealed by the following extract from Coase's (1960) work:

> In order to carry out a market transaction it is necessary to discover who it is that one wishes to deal with, to inform people that one wishes to deal and on what terms, to conduct negotiations leading up to a bargain, to draw up a contract, to undertake the inspection needed to make sure that the terms of the contract are being observed, and so on. [p. 15]

We will examine the relatedness of information and 'transaction' costs in greater detail in Chapter Six, however, this is clearly evident in selected works on the economics of search by Stigler (1961), on hierarchical control and optimum firm size by Williamson (1986b), and on information impactedness and vertical integration by Silver (1984).

Of immediate interest to our broader macroeconomic analysis of information occupations is the ambitious accounting exercise carried out by Wallis and North (1986). The latter have attempted, notwithstanding the 'fuzziness' surrounding transaction costs as an empirical category (see Hodgson, 1988), to measure the size and growth of the monetised 'transaction sector' in the US economy for the 1870–1970 period (Table 2.12). Given a high degree of overlap with our taxonomy of 'information occupations', Wallis and North's measured trends in

Table 2.12 *Employment in transaction-related occupations as a percentage of total employment, by industry, 1910–70*

Occupation	1970	1960	1950	1940	1930	1910
All employment						
With military	37.29	32.45	30.98	28.13	26.02	17.45
Without military	38.78	33.72	31.77	28.27	26.35	17.49
	Non-transaction Industries					
Agriculture,						
forestry, & fisheries	3.75	1.92	5.05	0.65	2.05	0.51
Mining	25.40	21.03	10.81	11.80	8.79	5.95
Construction	20.32	17.72	15.72	11.48	9.45	1.41
Manufacturing	30.22	27.88	24.30	22.22	19.27	12.53
Transportation,	37.62	37.43	33.63	36.44	32.46	28.29
communications,						
& utilities						
Services	28.09	23.09	19.78	12.46	12.70	5.40
Government						
With military	28.53	26.17	30.11	42.90	36.69	37.92
Without military	38.53	37.46	42.88	46.40	38.71	40.38
NEC	–	2.62	14.14	29.56	24.00	–
	Transaction Industries					
Retail trade	57.54	59.85	64.12	65.21	85.74	86.41
Wholesale trade	63.59	67.06				
FIRE	92.02	88.51	84.34	83.04	93.69	98.94

Note: Transaction-related occupations include: accountants; lawyers and judges; personnel and labour relations; farm managers; managers; clerical; salesworkers; foremen; inspectors; guards and police.

Source: Wallis and North (1986), p. 95.

so-called 'transaction-related' occupations are strongly, but not perfectly, correlated with the US 'information sector' trends presented earlier (see Table 2.1). We can, therefore, conclude that transaction and information costs are closely related as Williamson and other 'institutional' economists suggest — and postpone our theoretical exploration of this association to Chapter Six.

Let us now concentrate on the informational dimensions of the so-called 'new service' economy (Gershuny and Miles, 1983; Stanback, Bearse, Noyelle and Karasek, 1981). For this purpose, I have

Table 2.13 *The information component of Canada's labour force by Singlemann's industrial classification, 1971–81*

SECTOR AND INDUSTRY	Change in Total Labour Force (per cent) (1971–81)	Information Workers in Total Labour Force (per cent) 1971	1981	Diff.
AGRICULTURE, EXTRACTIVE, & TRANSFORMATIVE INDUSTRIES				
Agriculture	– 0.7	6.0	13.4	7.4
Extractive & Transformative				
Mining	52.8	34.7	42.4	7.7
Construction	39.7	23.7	30.2	6.5
Manufacturing	29.9	38.2	39.3	1.1
SERVICES				
Distributive services				
Transportation, communications, other utilities	39.9	50.2	53.8	3.6
Wholesale trade	61.0	61.7	62.3	0.6
Retail Services	52.0	42.8	45.5	2.7
Non-profit services				
Health & welfare	66.9	32.3	35.4	3.1
Education	35.8	84.5	86.0	1.5
Religion	30.4	21.8	23.2	1.4
Producer services				
Finance, insurance & real estate	74.6	91.7	94.0	2.3
Business services	130.2	85.8	86.0	0.2
Mainly Consumer services				
Amusement and recreation	76.2	39.8	44.8	5.0
Personal services	2.0	9.6	12.0	2.4
Accommodation and food	96.6	21.1	23.7	2.6
Miscellaneous services	109.5	47.8	50.3	2.5
Public Administration	38.8	54.2	62.1	7.9

Source: Hepworth (1987a)

reclassified the Canadian information occupational data (Hepworth, 1987a) by Singlemann's (1978) taxonomy which differentiates service sector activities by their function and market characteristics (Table 2.13). The occupational composition of the fast-growing 'Producer Service' industries clearly indicates that their outputs are information-related services, which mostly figure as intermediate inputs to the final products of all other industries. Their economic incidence does, in addition, extend to equivalent information services — accountancy, personnel, advertising and other activities — which are produced and used as 'in-house' resources within other sectors. The growth of this 'in-house' producer services component (in the Canadian economy) is evident from the large number and increasing proportion of 'information workers' in manufacturing, as well as other sectors.

The 'newness' of the service economy in Canada, as Table 2.13 indicates, has a second dimension, which is consumption rather than production related. The substantial growth of the 'Consumer Services' labour force has largely been concentrated in 'non-information' occupations, characterised by low-wage, part-time and female-dominated employment — for example, fast food workers, hotel attendants and household servants (Hepworth, 1987a). We have supporting evidence of these dual information-production/non-information-consumption dimensions of the 'new' service economy for other countries (Cohen and Zysman, 1987; Rajan and Pearson 1986). These dual dimensions are, for example, highlighted by Castells (1987), who refers to information-based production as 'high technology production. According to Castells:

> The development of information-based production is not limited to manufacturing: it actually requires a concomitant expansion of so-called 'producer services', so that the distinction between goods-producing and services-producing industries is increasingly blurred. The productivity generated by high technology and the income it induces in the highly-paid professional sectors create the basis (both in terms of investment and demand) for a wide range of service activities and consumer-oriented downgraded manufacturing, fed by labour displaced from other sectors or other countries. Both processes reinforce each other and require each other. So polarisation of the labour force is a fundamental and necessary process of the informational mode of development. [p. 47]

The future of information occupations

The reported slow-down of the occupational information sector's rate of growth in some industrialised countries (OECD, 1986), is consistent with Jonscher's (1983) predictions for the US. The latter forecasts a levelling-off and post-1980 decline in the sector's share of the US labour due to the differential impact of innovations in computer-telecommunications technologies on labour productivity in information and non-information occupations (see Table 2.14). A more complex view of the future occupational profile of the labour force is provided by Miles (1988a), as follows:

> While it is dangerous to extrapolate from current experience with IT systems to new generations of hardware and software, evidence to date indicates an association between IT-based change and some job loss, though less than thought likely at the end of the 1970s; an upgrading of some jobs and a demand for new technical skills and skill combinations; but also the degradation of other jobs, and a shift in the conditions of work for 'peripheral workers'. It is appropriate to speak of a *polarisation* of the workforce. [p. 11]

We are, in fact, witnessing an extremely complex set of occupational changes in the labour force rather than Jonscher's (and the OECD's) one-dimensional, 'levelling-off' of information occupations. Within the information sector itself, the share of professional, scientific and technical occupations is increasing, but the share of routine information-handling clerical, administrative and support staff in most industries is decreasing. Further, although the relative importance of unskilled manual occupations and single-skilled

Table 2.14 *Productivity trends in the information and non-information sector of the US labour force, 1950–2000*

| | Per cent | | | | | |
	1950	1960	1970	1980	1990	2000
Annual productivity growth in information jobs	0.5	0.6	0.8	1.3	2.3	5.1
Annual productivity growth in non-information jobs	4.6	4.7	4.0	1.0	0.7	0.6
Information occupations in the total labour force	39.7	44.5	48.7	50.5	49.5	45.7

Source: Jonscher (1983), p. 34.

craftsmen in all sectors is declining, this contrasts with the ascendancy of multi-skilled craftsmen, particularly in manufacturing industry (Rajan and Pearson, 1986). In addition to these occupational changes, 1980s trends in the UK, at least, indicate significant changes in the organisational locus and contractual basis of employment, including: the relative growth of self-employment, temporary and part-time working, double-job holding, full- and part-time female labour and small business employment (see Johnson, 1988).

On the basis of these trends, Hakim (1987) classifies the British workforce into a 'stable' (core) sector, made up of full-time employees, and a 'flexible' (peripheral) sector, comprising part-time, temporary and self-employed workers. He estimates that between 1981 and 1986 the 'flexible' sector of employment increased from 30 to 35 per cent; 'flexible' occupations tend to be mostly sales, catering, construction, literary, artistic and sport employment, with the main industrial concentration of labour being in agriculture and construction, distribution, hotels, catering and other services; over half of all female employment is describable as 'flexible'.

Very clearly, these recent trends in the labour market point to fundamental changes in the general economic and institutional contexts in which information occupation developments are now occurring. The variegated employment relation in so-called 'flexible' labour markets (see, for example, Piore, 1986) contrasts most obviously with the 'rigid', long-term contractual arrangements of the private and public bureaucracies, which both Porat (1977) and Beniger (1986) identify as the main source of information employment growth through much of this century. Thus, if Piore and Sabel (1984) are correct in seeing a genuine end to these bureaucratic systems of 'Fordist' regulation with the latest (1970s) capitalist 'crisis of control', then what they call 'flexible specialisation' will undoubtedly have a profound impact on the shape of the occupational information sector.

These are, however, the early days of contemporary 'flexible specialisation' for the bulk of manufacturing, notwithstanding some retroactive labelling of more long-standing, regional complexes of small firm industrial activity (see, for example, Storper and Christopherson, 1987). It is not obvious what the net effects on the information sector of these 'neo-Fordist' (Aglietta, 1979) or 'post-Fordist' (Roobeek, 1987) developments in advanced capitalism are or likely to be. According to Strassman (1985), for example, the 'replacement of the firm by the market' (vertical disintegration) should lead to a net increase in the size of the information (occupation) sector, given that the fraction of information workers declines as organisational size increases. The operation of these scale economies in information-handling (Arrow, 1984) is, however, contradicted by Scott's 1986a) evidence on vertical disintegration patterns in a regional complex of 'high-tech' manufacturing firms (Orange County, California). Scott observes, somewhat tautologically, that the transfer of work from 'plants with overgrown bureaucracies to plants with proportionately smaller managements [has resulted] in rapid increases in the proportion of blue-collar workers in the manufacturing labour force' (p. 150), but then adds that this 'runs absolutely counter to the trend for the United States as a whole' (p. 158).

Understandably, we do not know what a 'non-Fordist' information sector looks like. It would, in the event of 'markets replacing hierarchies' not only in private enterprise but also in public services (Aglietta, 1979), certainly have a different inter-regional and international geography. As I have remarked elsewhere (Hepworth, 1988b), tendencies towards economic decentralisation (from 'hierarchies to markets') may produce a more distributed 'information space' economy, contrasting with Pred's (1977) hierarchical, 'post-industrial' model of a metropolitan system dominated by corporate bureaucracies and information-impacted agglomerations of quaternary services activity (see also, Marshall, 1988).

Conclusion

The presented evidence on occupational change indicates that production processes in advanced industrialised economies are growing more and more information intensive. Recent extensions and refinements to this division of information labour have their historical origins in the middle-nineteenth century period of industrialisation. Throughout much of this century, the growth of informational activity has been concentrated in private and public bureaucracies, which developed as the key 'control technologies' of capitalist economies based on mass production, mass distribution and mass consumption. In the last two decades, we have witnessed some signs of the 'break-up' of these bureaucracies, as a small but growing share of informational activities have been externalised to producer services markets.

However, if the numerous pundits of so called 'post-Fordism' prove to be right, we are likely to witness more profound changes in the information sector of the economy and workforce in capitalist societies. The materialisation of strong trends towards vertical disintegration in manufacturing and service industries, together with the 'rollback' of Keynesian demand management and the privatisation of public services, would act to raise the *economic demand for control* over production, distribution and markets. The general results would be, as with previous 'crises of control' (Beniger, 1986), to raise the information intensiveness of firms, industries and the economy as a whole. Importantly, the development of these informational resources used to deliver the control requirements of the capitalist economy is marked by *historical regularities and continuity*, as our secular trends in the informational labour force clearly reveal.

We noted the uneven development of the information economy in an inter-regional context, drawing on some available evidence from Canada and the UK. What is most striking about this evidence, to my mind, is the growing importance of informational activities in all regional economies of these countries, whether these regions are generally thought to be 'central' or 'peripheral'. The most plausible explanation for this apparent evaporation of centre–periphery relations is the 'blurring' of market and technological boundaries that separate manufacturing from services, which the recent phase of growth and specialisation in the information economy has brought about. The information economy is, however, not a 'leveller': what we are witnessing is the containment of centre–periphery relations within a more extensive and refined spatial division of information labour, which the new technologies are widely expected to dramatically reshape over the next decade and more.

Let us, then, turn our attention to the capital resources of the information economy, focusing on their latest and most powerful vintage — namely, the converging technologies of computers and telecommunications. For some observers, the 'information economy', as a new or qualitatively different stage of capitalist development, begins with the advent of these technological innovations (see, for example, Stonier, 1983; Miles, 1988a). The certain history of 'information occupations' (see Table 2.1) and Beniger's long list of 'control innovations' (see Figure 1.1) do not, however, suggest that we are witnessing a clean break with the past.

Note

1. Porat's work on 'The Information Economy' covers nine volumes, including background discussion of time-budget studies on the information content of various occupations.

Chapter Three
Information Technology as Spatial Systems

Introduction

Information technology is a generic term for the widening array of electronics-based products and services generated out of the '*convergence*' in computer and telecommunications innovations. As new elements of the capital stock, used by firms, governments or the household, these technologies are now transforming production techniques in all sectors, methods of management and forms of industrial organisation, the product composition and contours of markets and modes of consumption. In short, information technology has a fundamental influence on the economy and its institutional arrangements, accounting for its description by Freeman (1987) as a 'generic technology'.

In this chapter, I begin by conceptualising information technology as a distinctive form of capital, and then proceed to describe its spatial attributes. This exposition of information technology as spatial systems necessitates a brief technical account of how and why computer and telecommunications 'converge' to form capital combinations known as computer networks. It is this computer network model which constitutes the technological focus of the book's discussion.

Importantly, for modelling information technology, this chapter focuses on wide area computer networks (WANs) used as proprietary systems by large, multilocational firms and governments. Further, it does not elaborate the local configuration of these WANs at specific workplaces — that is, the industrial local area computer networks (LANs) used in factory automation (flexible manufacturing systems) or office automation systems. A complete specification of technological change would, therefore, require consideration of WANs and

LANS and their interconnection, together with inter-organisational computer network innovations (e.g. 'just-in-time' technologies) where appropriate. In this respect, several concepts, such as computer network topology and architecture, are common to different levels of 'networking', and may, therefore, be useful for specifying technological change and innovation 'barriers'.

Information capital

Capital is not homogeneous (or malleable) as it is conceived in traditional Keynesian and neo-classical models of production and distribution. The heterogeneity of capital resources, as everybody knows now, poses 'measurement problems' for economists and geographers alike. (How do we add computers to trucks?) Despite these 'problems', which are usually 'solved' by adopting money value as a standard measure of the size of capital stocks and investment volumes, we clearly cannot do without a generic concept of capital. Thus, geographers have tended to 'homogenise' capital resources in empirical studies, while adding as caveats the familiar neo-Keynesian objections to the neo-classical treatment of capital as a homogeneous and malleable factor of production (see, for example, Gertler, 1984, 1986a).

The capital resources used by the firm are heterogeneous in terms of use. Their joint or complementary use, as capital combinations, form a pattern which Lachmann (1978) calls the 'capital structure'. This conception of capital is, according to Lachmann, that of 'a complex structure which is *functionally differentiated* in that the various capital resources of which it is composed have different functions' (p. 7). His rationale for treating the capital stock as a structural pattern, rather than a homogeneous aggregate, is stated as follows:

> A theory of investment based on the assumption of a homogeneous and quantifiable capital stock is bound to ignore important features of reality. Owing to its very character it can only deal with quantitative capital change, investment and disinvestment. It cannot deal with *changes in the composition of the stock*. Yet there can be little doubt that such changes in the composition of the stock are of fundamental importance in many respects, but in particular with regard to the causes and effects of investment. [p. 6]

The past decade, of course, has witnessed major changes in the composition of the capital stock at the level of the firm and of the economy. It is generally accepted that, by far, the most important new elements of the capital stock are computer and advanced telecommunications innovations — the new, electronics-based information technology (Jonscher, 1983). These powerful innovations have prompted Porat (1978) to suggest revisions to the simple, neo-classical production function model in a way that is consistent with Lachmann's conception of a functionally differentiated capital structure.

It is customary for orthodox economists and economic geographers to model production as a stylised process which can be reduced to the following mathematical expression: $Y = F(K,L)$, where Y is output, L is labour and K is capital. That is to say, no account is taken of how changes in output (growth) arise from changes in the composition of either the capital stock or the labour force. For example, computers and chain saws are lumped together as capital; similarly, scientists and carpenters are treated as indistinguishable components of a single, homogeneous, category of labour.

The revisions to this traditional production function that Porat (1978) advances constitute an attempt specifically to separate the output contributions of information and non-information resources. He suggests that, for capturing the separate effects of the new information technology, the model should be revised to take on the following expression:

$$Y = F\ (K_I, K_N, L_I, L_N)$$

where Y = output

K_I = information capital inputs (e.g. computers)
K_N = non-information capital inputs (e.g. lathes)
L_I = information labour inputs (e.g. clerks)
L_N = non-information labour inputs (e.g. machinists)

The examples of the dual capital components, presented in brackets, are those offered by Porat (1978: 21).

Conceptually, the differences between information and non-information capital is that the former types of capital are primarily used to produce, process and distribute information (through some medium); the latter are allocated directly to the creation and circulation of physical resources. For example, blast furnaces and trucks are used by the steel industry to manufacture metal sheets and transport them to markets; the same industry also uses telephones and computers to transmit and process information in support of these activities — the scheduling of shipments, maintenance of accounts payable and so on. Porat (1977), and other researchers applying the 'four-factor' model (for example, Warskett, 1981), rely on an inductive definition of information capital, by classifying the output of numerous goods-producing industries (computer manufacturing, telecommunications equipment, office machinery, etc.) according to a broadly stated criterion: the good (as an intermediate production input) must 'intrinsically convey information or be directly useful in producing, processing or distributing information' (Porat, 1977): 25).

The rationale for differentiating capital inputs derives from Porat's explicit concern with isolating the economic role of the *new* information technology. This is clearly suggested by his chosen emphasis in describing the capital inventory of the individual firm:

> A wide variety of information capital resources are used to deliver the informational requirements of one organisation: typewriters, calculators, copiers, computers, telephones and switchboards. And, depending on the size of the organisation, there could be a massive array of *high technology* information goods such as microwave antennae, satellite dishes, and facsimile machines. [1977: 2]

The added emphasis on 'high technology' is my own doing. It is intended to underscore the basic intent of the 'four-factor' model, with respect to its functional differentiation of capital resources: that is, the new vintages of information capital goods are so much more superior (in terms of marginal efficiency) to their mechanical antecedents, that fundamental changes in received models of production are warranted to capture their unique and separate contributions to total factor productivity.

The Porat model is still on the 'drawing board'. Further discussion of the model is reserved for Chapter Four, where we will explore its elaboration in a spatial context more thoroughly. It should be noted even now, however, that several basic objections can be levelled at the model's formulation and recommended mode of operationalisation. First, and perhaps most obviously, pure information is left out of

Porat's revised production function — information and knowledge are still needed to specify factor intensity in terms of capital–labour combinations. As pointed out by Arrow (1984: 170): 'A key characteristic of information costs is that they are in part capital costs; more specifically, they typically represent an irreversible investment.'

Second, by lumping together all forms of information technology and measuring information capital inputs in terms of inter-industry transactions, the model cannot account for the synergistic effects of technological convergence in computers and advanced telecommunications. It is, however, generally recognised that the powerful capabilities of these innovations for transforming production processes derive from this 'synergy'.

And third, the procedure of classifying capital inputs will become increasingly difficult as certain forms of 'non-information capital' becoming increasingly 'intelligent'. Examples of these ambiguous capital components – compare Porat's '50:50' occupational categories of 'information labour' — are robots and computerised machine tools, which are clearly allocated to direct production tasks, although they 'intrinsically convey information' and are 'directly useful in producing, processing or distributing information' (Porat's criteria for classifying capital as informational, as previously stated). As such, separating out the informational and non-informational components of capital, in manufacturing industry particularly, is not an unproblematic task. This is because of the firm's increasingly computerised capital structure and its attendant information structure.

Regardless of these objections to Porat's model, I propose to retain the analytical category of information capital because, as a generic concept, it is an illuminating and empirically useful one. For example, in the following chapter, the concept proves useful for highlighting the spatial structure of the new information technology and its profound implications for modelling the inter-regional dynamics of capital. We will see that the new technology can not be modelled as though it were just like any other form of capital — as Porat (1977, 1978) assumes. Also, it becomes clear that 'computer network' technology, as a form of information capital, poses some new and different problems for regional economic geographers to manage.

For the present, attention will focus on the manner in which the new information technology is organised into spatial systems. This involves an exploration of the technical characteristics of computer and telecommunications innovations, and a brief preamble.

With the proliferation of computers and terminals and their widening geographical distribution, over the last two decades, the recent thrust of technical research has been towards making different machines at different locations work together. During the 1970s, this spatial 'division of capital' (Lachmann 1978), arising from the functional specialisation of machines as electronic information capital accumulated, came to be implemented in the form of wide area 'computer networks'. As a result of this spatial division of information capital, enabled by the technical capabilities of computer devices for inter-communication over vast distances, individual firms are now able to achieve greater flexibility in production organisation. We will turn to some of the economics of this 'flexibility' in the last section, after a necessary technical exposition of computer networking.

Computer networks

The term computer network is normally used in a generic sense and various definitions appear in the technical literature:

> A computer network is an arrangement of computer systems and the facilities needed to access them and store information. The various elements are linked by some means, usually telecommunications circuits [Gee, 1982: 29]

> The old model of a single computer serving all of the organisation's computational needs is rapidly being replaced by one in which a large number of separate but interconnected computers do the job. These systems are called computer networks. [Tanenbaum, 1981: 2]

There are many types of private wide area networks in existence (see Chapter Five). These are proprietary systems used exclusively by one organisation to link together computers and terminals at several remote sites by means of telecommunications circuits. Private networks may also be closed systems shared by a community of authorised end users — for example, the SWIFT (Society for Worldwide Interbank Financial Transactions) and SITA (Société Internationale de Télécommunications Aéronautique) networks used by the international banking and airline industries, respectively. Public networks, by way of contrast, are operated by telecommunications carriers as a universal service, and local area networks (LANs) are private systems confined to one site, such as an office building a or university campus.

The term 'end user' requires clarification. It is often defined technically as any intelligent process which utilises a computer network, and generally refers to an application program in a computer or terminal. Here, the terminology preferred by Martin (1981: 143–4) is adopted: the end user is the organisation which uses the network system and the user may be an application program, the human operator of a terminal, or any mechanism which is the immediate source or sink of information transmitted over the network.

Network components

A private computer network is made up of the end user's machines and transmission facilities obtained from telecommunications carriers. The system of capital, built with this information technology, is used for: transporting data between users and supporting meaningful communication between them; and processing information to carry out applications.

As an analogy, a network of information workers would consist of the workers themselves — as 'processors' — and the telephone sets and lines through which they communicate. Telephone conversations between information workers are voice communication, whereas the encoded messages exchanged between machines in a computer network are referred to as data communication. A computer network may carry digitised voice and data together.

The end user's machines specialise in two basic types of functions:

(*1*) data processing, or simply computing, is the activity of processing information to carry out particular applications, such as data analysis — the information content of data is changed;

(*2*) communications processing, is the activity of transporting messages between users and supporting meaningful communication between them — the information content of data is assumed to be unchanged *en route*.

On the basis of their functional specialisation, the end user's machines occupy different tiers of a network hierarchy. Further, because machines are distributed across space, a computer network is essentially a system for processing and distributing information implemented as a spatial hierarchy of information capital.

The transmission facilities in a private network consist of: telecommunications circuits, or the channels through which data are transported between users; and user/network interface devices, or devices for linking the end user's machines to the carrier's network. The characteristics of telecommunications links differ. Data are transported over various physical media (microwave radio, coaxial cable, satellite channels, wire pairs, or fibre optic cable), using analogue or digital transmission techniques, and routed by different switching systems. Links also differ in terms of information-carrying capacity or bandwidth, and may be private or public telecommunications circuits.

Where computer-generated data are transmitted over the ordinary telephone system, an interface device is needed for converting digital signals to, and from, analogue form. This device is called a modem (modulator/demodulator) and its 'translation' role is illustrated in Figure 3.1. Modems may also handle the initial set-up of a 'conversation' between machines co-operating to process an application — an exchange of control messages known fittingly as 'handshaking'. Modems are not needed for digital networks, although some other device may carry out similar interfacing functions — for example, PAD machines (Packet Assembler/Disassembler) are used for interfacing start–stop, character mode terminals to a packet-switched network.

Some basic components of a network appear in Figure 3.2. The

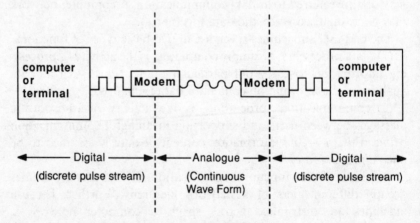

Figure 3.1 *The signal translation role of modems*

Physical layout

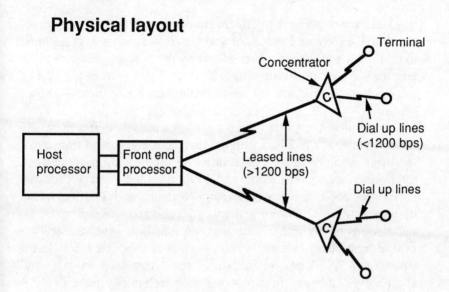

Functional hierarchy

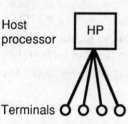

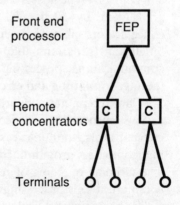

Figure 3.2 *Basic components of a computer network*

Source: Hepworth (1987a)

machines which primarily perform communications-processing functions are the front end processor and the concentrators. It is assumed that all data processing is carried out by the host processor and the terminals are simple input–output devices. The hierarchical order of the machines, as defined by their functional specialisation, is shown below.

Communications processing is a complex activity that is normally transparent to, or hidden from, the user. The end-to-end transfer of messages involves various procedures — establishing user/network interfaces, message switching and routing, polling and addressing, speed and code conversion, message formatting, buffering, queue management, error checking, diagnostics and record-keeping. Originally, nearly all of these communications functions were handled by a central computer system, which was primarily used for batch processing. As computer networks have increased in size and complexity, advances in semi-conductor technology have made it feasible to reallocate, or 'off-load' many of these functions from an all-purpose mainframe machine to several, specialised, machines at remote locations.

In Figure 3.2, a specialised computer called a front end processor handles the interface between the host processor and the public network, and assumes responsibility for overall network management and control — the host is released for data processing functions. They are linked by a parallel computer-to-computer interface. The concentrators play an intermediate role at remote locations, and 'intelligent' (programmable) terminals carry out limited but important communications processing functions at the user sites — such as message formatting and error checking.

A large computer network is configured with numerous communications processing devices, including front end processors, concentrators, multiplexors, terminal cluster control units, and minicomputers programmed for communications-oriented functions, The ultimate outcome of recent technological developments is that communications processing is now a spatially distributed activity, as well as being hierarchically organised. Data processing has undergone similar structural changes. In Figure 3.2, it is assumed that no applications are processed locally by the terminals. Today, the buffered storage of intelligent terminals enables users to draw on remote data bases and process information locally using standard

HOSTS DISTRIBUTED BETWEEN CITIES

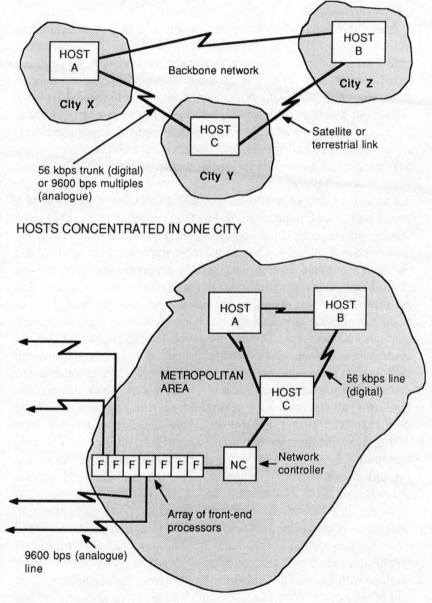

HOSTS CONCENTRATED IN ONE CITY

Figure 3.3 *Computer networking with mainframes*

Source: Hepworth (1987a)

application software. With the advent of distributed processing, the spatial reallocation of data processing functions has assumed even greater dimensions.

A multiple host network is illustrated in Figure 3.3, in which the workload is shared by three mainframe computers — the tasks of data base management, applications processing and network control are distributed. Instead of being located in different cities, the hosts may be concentrated in one metropolitan area, and interconnected by very high-speed data lines (56 kilobits per second). In this case, a specialised, large computer may be dedicated to overall network control 'behind' an array of front end processors. These types of distributed systems have evolved as the computing requirements of large multilocational organisations have increased.

Data processing may also be spatially distributed in the form of a 'mesh' network of minicomputers located in different cities. The use of 'minis' with a mainframe computer in a hierarchical network is illustrated in Figure 3.4. The 'minis' serve the immediate needs of the end user's regional operations, such as processing applications for payroll and sales tracking. At the end user's headquarters, the mainframe machine is used to maintain a central data base, process organisation-wide applications (such as electronic mail or quality control) and develop standard programs for the regions. Again, in contrast with early network systems dominated by a central computing facility, data processing has evolved as a spatially distributed activity.

The telecommunications links in Figure 3.2 are analogue circuits. Dial-up analogue lines have a limited bandwidth because they have been engineered primarily for voice communications — they are part of the public telephone system used by household subscribers. Data are transmitted between users over a physical (copper) path established through circuit-switching techniques. The path is maintained only for the duration of the 'call' and the 'speed limit' for data transmission is 1200 bits per second. In contrast, private lines are point-to-point, permanent connections between the end user's machines. As dedicated non-switched circuits, they can be 'conditioned' (by means of equalisation devices) to carry data at speeds of up to 9600 bits per second, with the use of sophisticated modems. Private networks are still largely made up of leased circuits which can be split into several communication channels by multiplexing and concentration techniques.

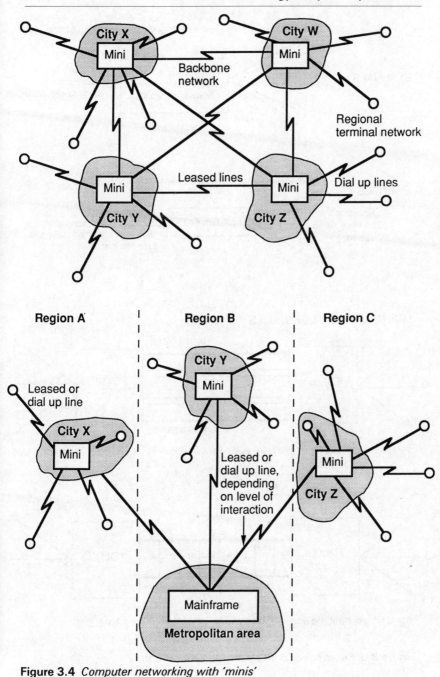

Figure 3.4 *Computer networking with 'minis'*
Source: Hepworth (1987a)

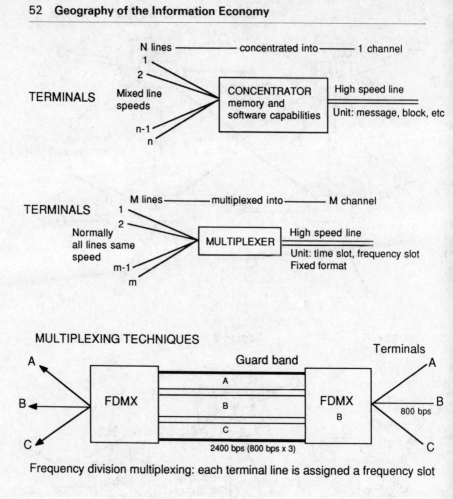

Frequency division multiplexing: each terminal line is assigned a frequency slot

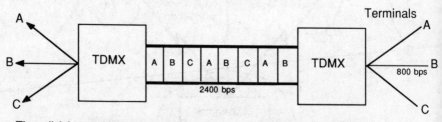

Time division multiplexing: each terminal line is assigned a time slot

Figure 3.5 *Functions of multiplexers and concentrators*

Source: Hepworth (1987a)

In configuring a network to minimise line costs, various machines are used to handle the typical 'burst' (intermittent) traffic generated by a large, dispersed population of computers and terminals. The buffer storage and programs in a concentrator enable it to store and forward messages exchanged between the host processor and the remote terminals. Like the postal system (speeded up), messages from different users are collected, bundled and transmitted together on a single telecommunications line — in Figure 3.2, terminal traffic is 'concentrated' onto two high-speed lines and separate host–terminal links are not required. Figure 3.5 illustrates how a concentrator carries out these functions, in relation to a multiplexer, which is also widely used to economise on line costs. (Other devices include terminal cluster control units and multidrop lines.)

Digital circuits are also used on a private line or dial-up basis. In Figure 3.3, these may be used as inter- or intra-city connections between the hosts — to support 'bulky' applications such as file transfer. In a packet-switched digital network, switching and transmission facilities are shared by all users. However, unlike the circuit-switched telephone system, data are routed dynamically between users over a virtual (logical) path selected by switches that store and forward messages. In contrast to message-switched networks, which are also based on store and forward technology, the user's messages are not transmitted as a continuous block of data. Instead, messages are fragmented into discrete units called packets, which are addressed, routed individually, and reassembled at the destination where messages are reconstituted or 'de-packetised'.

The current trend in the telecommunications carriage industry is towards the provision of digital circuits which can carry data, voice and image traffic together. These integrated services digital networks (or ISDNs) are already operated by satellite carriers which exploit a very high bandwidth to offer a mix of services from voice communication to teleconferencing.

Whereas public data networks are basically transportation systems, private networks are designed not only for transporting data between users but also for supporting meaningful communication between them. Private networks have evolved as systems built with a diversity of equipment supplied by a relatively small number of computer systems manufacturers. As networks have become complex, computer vendors (such as International Business Machines or Digital Equipment Corporation) have established their own methods for network configuration and control. These methods are set out formally in their network architectures, which are 'master plans' for distributed processing based on the vendors' own hardware and software products, incorporating a functional description of network components and the rules that govern their interaction.

Network architecture

A computer network is said to have a 'division of capital' — analogous to a division of labour — because machines can communicate to perform complementary tasks. When machines exchange information, or 'speak' to each other, they observe certain conventions or protocols which are similar to those used in human communication. For example, a telephone conversation between information works usually begins with the parties identifying themselves, and perhaps a confirmation that they can hear each other clearly. Following this initial exchange, individuals talk in an orderly fashion about a mutually recognised subject, and finally both agree to end the call and sign off. Thus, the exchange of information is subject to a finite set of protocols (including rules for acting on a 'breakdown' — redialling or 'repeat that' which control the conversation. Similarly, business letters usually include an opening phrase such as 'With reference to your letter dated …' to establish a relation between the original message and the response.

In computer networking, the transfer of data between machines is also governed by a set of protocols, as illustrated in Figure 3.6. Further, in the same way that people communicate by virtue of a common understanding of language and concepts, the messages exchanged between machines must have precise formats if they are to be intelligible to a variety of devices and acted upon. These protocols and message formats are rigorously defined in network architectures.

It is helpful to illustrate how computer-based systems communicate by means of an analogy. Let us assume that two scientists are exchanging the results of research directed at finding a cure for a fatal disease. One lives in Moscow and speaks only Russian; the other lives in New York and speaks only English. Thus, the scientists are separated by distance and language barriers, and to co-operate they require the additional services of translators and engineers. In Figure

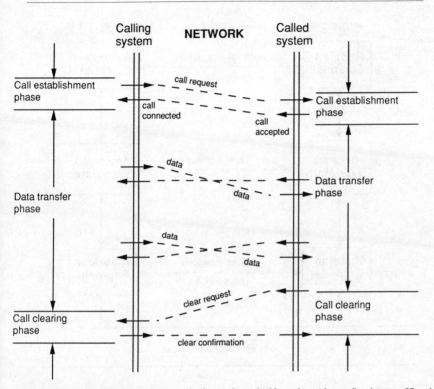

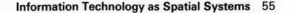

Note: Data and control messages have a precise format. A standard frame format for sending data on x.25 packet networks is shown below:

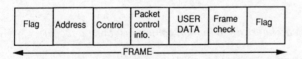

Flag	Address	Control	Packet control info.	USER DATA	Frame check	Flag

◄───────────── FRAME ─────────────►

Figure 3.6 *Conversations between computer systems*

Sources: R. Sharma, P. de Sousa, and A. Ingle (1982) *Network Systems* New York: Van Nostrand Reinhold, pp. 305–7.

3.7 this division of labour is modelled as a layered hierarchy.

To avoid talking at 'cross purposes', the scientists must begin by settling on an appropriate topic of conversation (perhaps, an ongoing conversation) — this is formalised in the 'layer 3 protocol'. For the American scientist to communicate with his remote counterpart, he

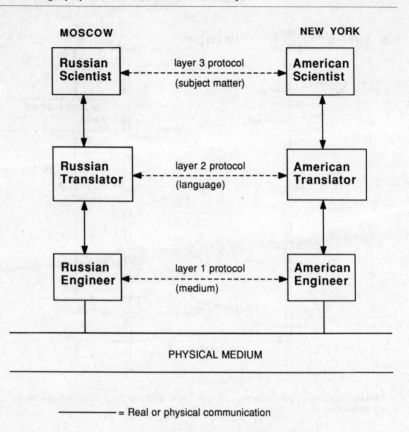

Figure 3.7 *The architecture of human communication (analogy)*

Source: Adapted from A. Tanenbaum (1981).

must pass his message (ideas) 'downwards' to a local translator. The translator converts the message into esperanto, or any other language which he and his Muscovite peer have agreed to use — this is formalised in the 'layer 2 protocol'. In turn, the translated text is forwarded to an engineer in New York, who transmits it as a string of electrical signals over a physical medium which he and his Russian counterpart have agreed on — the subject of the 'layer 1 protocol'. In Moscow, the process is reversed and the message travels 'upwards' through the layers: the Russian engineer receives the signals, and converts them to an intelligible form for the local translator; and, finally, the translated message is passed on to the Russian scientist to be 'processed'

Three basic aspects of this human model are relevant to inter-machine communication in a computer network. First, protocols must be established in advance — the scientists have to decide what to talk about; the translators have to choose a common language, and the engineers have to agree on a transmission technique. Second, each person thinks of his conversation as being horizontal — with his peer — although inter-personal contact is actually vertical. Each scientist need not concern himself with the details of message transmission and translation — these services are hidden or transparent. Third, this communication structure is modular — the scientists can switch subjects, the translators can switch common languages, and the engineers can switch transmission media — but, a switch in any one layer does not require a corresponding switch in any other layer.

In this analogy, a communication system is set up to enable two scientists to share information with the ultimate aim of developing a cure for a disease. A system with a similar 'wedding cake' structure is used in computer networking so that users can co-operate to process information in carrying out applications. As stated earlier, users may be application programs or human operators instead of 'scientists'; and communications support services are provided by sets of functions or peer processes ('translators and engineers') implemented in the form of specialised computer hardware and software modules. It is these peer processes that observe the protocols referred to earlier.

In computer networking, these mutually dependent processes are also modelled as a hierarchy of layers, and the functions they must carry out on behalf of the users are described intuitively as follows:

What does the user look like? (the Presentation Layer)
Who is the user? (the Session Layer)
Where is the layer? (the Transport Layer)
Along which route do we get there? (the Network Layer)
How do we make each step in that route? (the Link Layer)
How do we use the medium for that step? (the Physical Layer)

[Meijer and Peeters, 1982: 13]

The layer names are used in the Reference Model of Open Systems

Interconnection (OSI), as defined by the International Standards Organization. This model has been developed as a conceptual tool to identify areas in which common technical standards can be established to allow end users to interconnect equipment designed and built by

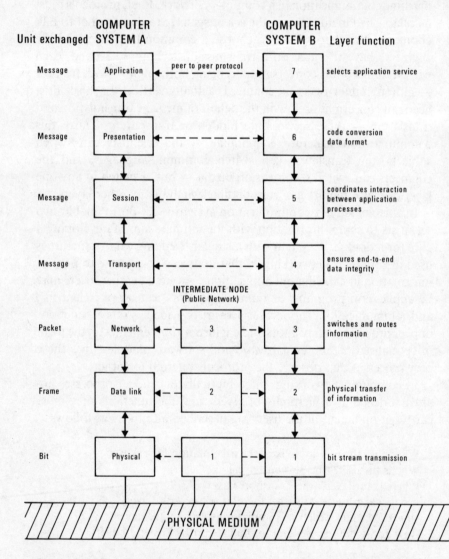

Figure 3.8 *The open systems interconnection model*

Source: International Standards Organization

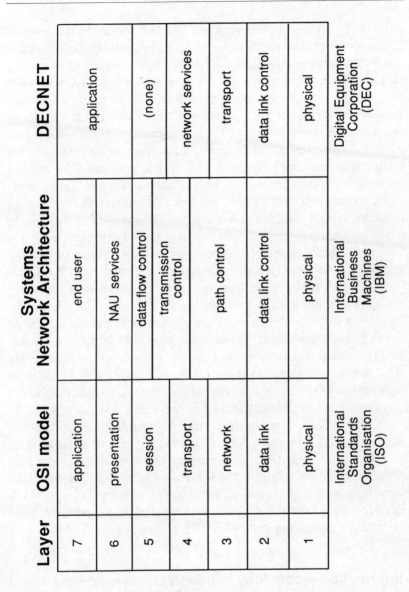

Layer	OSI model	Systems Network Architecture	DECNET
7	application	end user	application
6	presentation	NAU services	
5	session	data flow control	(none)
4	transport	transmission control	network services
3	network	path control	transport
2	data link	data link control	data link control
1	physical	physical	physical
	International Standards Organisation (ISO)	International Business Machines (IBM)	Digital Equipment Corporation (DEC)

Figure 3.9 *Selected network architectures of computer manufacturers*

Source: A. Tanenbaum (1981)

different computer systems vendors. The OSI Model defines a set of standardised procedures for the exchange of information between computer-based systems. These standards define the sets of functions, or peer processes, that should reside in equivalent layers of each system, and the peer protocols that should be observed in a set of parallel 'conversations' between the processes. The formal functions of the layers are shown in Figure 3.8. In the model, the user's message is passed 'downwards' across layer interfaces which define the functional relationship between adjacent layers, and is 'enveloped' by control information being exchanged between equivalent layers. These 'envelopes', which take the form of headers and trailers, are 'opened' in sequence in each layer as the message travels 'upwards' to the user. In other words, the frame of data that is physically transmitted is the user's original message or data accompanied by a series of supporting messages sent by the layers that act on the user's behalf (see Figure 3.6).

The OSI Model does not, however, specifiy the content of protocols, nor does it imply what technology should be used to implement the layers — it indicates only where common protocol standards could be developed. Proprietary architectures developed by individual computer system manufacturers do not use the same layers or formats for data and control messages (Figure 3.9). As a result of these different technical standards, network architectures are said to be incompatible — or machines of different make can not 'speak' to each other.

In sum, private computer networks have undergone a process of structural transformation. New technologies, such as concentrators and data communications, have enabled end users to distribute data processing and communications processing activity across numerous locations. In the next section, the concept of network topology is introduced as a formal description of the spatial characteristics of these systems for processing and distributing information.

Information technology in geographical space

The topology of a computer network is a representation of the geographical layout of its computer and telecommunications components. Network topologies are likened to various geometric forms — star, mesh, tree, ring — and resemble graph-theoretic descriptions of other real world distribution systems, such as railroads and energy grids. A large composite system may include interconnected sub-networks of various configurations, particularly if it is implemented with equipment purchased from different computer manufacturers.

Star   Tree   Parallel Tree

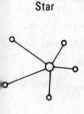
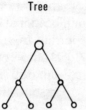
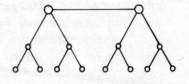

Mesh   Loop

Bus   Broadcast

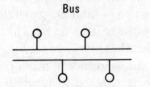
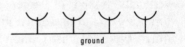

Note: The above figures are logical (functionally defined) representations of network topologies. Circles are network nodes (computers/terminals) and lines are communications links

Y Earth station

Figure 3.10 *Common network topologies*

Source: Hepworth (1987a)

Several types of topologies are presented in Figure 3.10. The central node of a star network — typical to service bureaux and retail chains — is normally a large general-purpose computer and peripheral nodes are terminal locations. In order to minimise costs of data communications, this type of centralised system is nearly always implemented with specialised devices for sharing transmission facilities — such as multiplexers, concentrators and terminal control units. Tree networks are generally used by multilocational organisations whose internal systems of management reporting are similarly structured — for example, the national/regional/local offices of a chartered bank — and hierarchies may be implemented with distributed minicomputers (see Figure 3.4), front end processors and also intelligent concentrators.

Most private networks are vertical structures and, with the advent of distributed processing, they may be interconnected at their higher tiers to form 'parallel-trees'. Horizontally structured networks, which imply that all components carry out equivalent processing functions (equal rank), are rarely implemented as inter-city systems. Only satellite broadcast networks fall into this category, although signals are normally transmitted from a single central node. Bus and ring topologies are implemented only in local area networks which are cable-based systems interconnecting user equipment at a specific site. Finally, mesh networks (see Figure 3.4), in which data and communications processing functions are evenly distributed between several minicomputers, tend to be used for specialised and regionalised functions — for example, product distribution — and are generally vertically linked to larger integrated systems. The approach to topology design has much in common with those used to configure other types of distribution systems — transportation networks, electricity grids, etc. The design problem is reduced to creating a backbone network, linking the main switching nodes by high-capacity trunk lines, and local access sub-networks that link computers and terminals to the backbone by low-grade lines. The problem has been formulated by Tanenbaum (1981: 32–3) as follows:

Given:
   Location of the hosts and terminals (sources and sinks)
   Traffic matrix (volumes x types)
   Cost matrix (inter-nodal line costs by grade of service)

Performance constraints:
Reliability (connectivity)
Delay/Throughput (response times)

Variables:
Topology (location of lines and switching devices)
Line capacities
Flow management (routing)

Goal:
Minimise cost (line and device costs)

Software packages for modelling and simulation are normally used in network design to optimise the physical layout of lines and a variety of communications-oriented devices, such as concentrators, front end processors, multiplexers, terminal control units and so on. The basic, dynamic, influences on the design process are the cost and capabilities of telecommunications and computer technologies. The topology of a private network will, however, be determined ultimately by the spatial organisation of the end user's basic operations and management strategy for using information technology. Political factors also affect network topology by imposing constraints on locational behaviour.

Network topologies are generally understood to be maps showing the types of information technology being used at different operations sites. When obtained directly from individual organisations, these maps may appear to be extremely technical documents which vary by levels of abstraction. In general, they will provide details on the wide-area *and local-area* configuration of computer networking — that is, the spatial division of information capital within and between operations sites. As stated earlier, this chapter focuses only on the 'backbone' topologies of wide area system.

Network topology receives relatively little attention in the technical literature, which is primarily concerned with architectural solutions to distributed processing and the roles of individual technologies. Further, the topological design problem is stated in a form which is generally applicable to all sorts of distribution systems. For our purposes, however, network topology is a key concept because computer networking can be situated in a spatial context. Quite literally, it allows us to put new forms of information capital — or information technology — 'on the map'. Further, the concept of network architecture is of central importance to understanding how electronic information capital is organised and used in practice. Specifically, computer networks are information technology as spatial systems characterised by a highly articulated division of capital. The case study research in Chapter Five applies this conceptualisation of information technology to analyse network innovations in large multilocational companies.

Some possible implications of the computer network model

The foregoing technical analysis has been particularly elaborate in its description of computer networks and their attendant spatial divisions of information capital. This close attention to detail is, however, necessary to effective geographical research on information technology, and, indeed, on any other important technology. Let us then consider what potentially useful insights the technical analysis has thrown up.

First, it underlines the symbiotic relationship between computer and telecommunications innovations, such that they must be treated as 'companion' or non-separable technologies in geographical research. We saw that the spatial diffusion of computer resources depends on the spread of telecommunications, and vice versa. The spatial distribution of information technology refers, therefore, to these functionally differentiated capital combinations; the inter-regional allocation of this 'networked' capital within firms (and between firms, in the case of inter-organisational computer networking) will reflect its characteristic implementation mode as resource-sharing systems, which offer users scale and scope economies in both computing and telecommunications.

Our 'network' model has obvious relevance for studies of technological diffusion in a spatial context. In particular, unlike other forms of productive capital, we cannot identify the spatial distribution of 'networked' information capital simply by counting or assigning a value to machines on a region-by-region basis (see, for example, Lesser, 1987). After all, computing resources can be 'remotely' accessed on a trans-regional basis, such that a regional 'tally' of the machine population *will* reveal nothing of the true level of local technological capacity. We would, therefore, need to take full account of the inter-regional flows of these information capital services in order to map out the spatial diffusion process. The broader implications of this 'communicability' characteristic of information capital for modelling the spatial dynamics of the economy are elaborated at greater length in Chapter Four.

A second line of research inquiry is suggested by the concept of network topology. It concerns the extent to which the topological characteristics of actual private networks, operated by multilocatioal organisation, have shaped and continue to shape the geography of public computer networks — the so-called 'electronic highways' of the information economy (Castell, 1985). This 'shaping' influence, which finds expression in the capital investment decisions of telecommunications carriers, acts through the differential effects on route profitability of the aggregate geography of data traffic configured by private network topologies. It has been highlighted by Schiller (1982) in aspatial terms, but received scant attention by geographers researching the role of telecommunications in regional development (see, for example, Goddard and Gillespie 1986; Pye and Lauder 1987).

We know for certain that large multilocational corporations and government agencies are the dominant users of advanced telecommunications services, and that it is their extensive, inter-city communications requirements which have driven the development of the 'electronic highways' in a national and an international context. In the UK, for instance, 60 per cent of all data communications traffic is generated by 300 large companies[1]; recent estimates for Norway indicate that only 25 companies account for about 50 per cent of the country's total data traffic.[2] Clearly, although this level of user concentration will vary between countries, the networking strategies of multilocational organisations must have a considerable influence on the spatial evolution of public networks and thereby relative accessibility to the new communications services.

We know very little (if anything at all) about the spatial cost surface of the information economy, when (partially) defined in terms of interurban or inter-regional price differentials for the new telecommunications services. What we do know, however, is that these price differentials are marked (Hepworth, 1986b; Gillespie, Goddard, Robinson, Smith and Thwaites, 1984), and that they enter into the cost calculus of private network planning and design. The cumulative effects of these cost influences on topological design, aggregated over a large number of private network users, would be to shape geographically the development of public networks. Thus, although this author has found that telecommunications expenditures in large Canadian firms may amount to less than 1 per cent of total operating costs (Hepworth, 1987a), the spatial externalities of corporate network planning decisions cannot be discounted as insignificant. These externalities, as expressed in relative accessibility to public telecommunications services, are clearly important to the locational behaviour of the so called 'on-line' industries (Howells, 1987) and to the inter-regional competitiveness of small firms (Gillespie and Williams, 1988). All in all, we are nowhere near the 'frictionless surface' which some geographers (see, for example, Abler, 1975) have predicted to materialise with the advent of advanced telecommunications systems; the 'even hand' of satellite broadcast communications, in this regard, applies only to the largest corporations (and then only to computer network applications which tolerate that medium's signal propagation delay).

The third set of research issues raised by the 'network' model relates to the economics of the multilocational firm. I will introduce some of this economic theory now and develop it further in following chapters. Of particular interest is the economies of scope which computer networks, as 'resource-sharing' systems, make available to the multilocational firm.

Economies of scope exist when for all outputs q_1 and q_2, the cost of joint production is less than the cost of producing each output separately (Panzar and Willig, 1981). This condition is normally expressed as:

$$c(q_1, q_2) < c(q_1, 0) + c(0, q_2)$$

From a neo-classical perspective, then, economies of scope are viewed as providing cost savings in joint production which are generally available to any multiproduct firm. They arise (according to Panzar and Willig, 1981), from the 'quasi-public' property of shareable inputs in production. The most obvious classes of common inputs which exhibit this public good property are specialised and indivisible physical assets and information (Teece, 1980).

Thus, in characterising computer networks as 'resource-sharing systems', we are more precisely referring to these indivisibilities associated with information (Radner, 1970) and with the utilisation of computing assets (hardware and software). It is the existence of these two types of indivisibilities which yield economies of scale and scope in production at the level of the firm. Further, as emphasised earlier, these economies are realised and take effect over space: 'computer networks are information technology as spatial systems'.

The importance of information services or 'intangible assets' is closely related to the concept of economies of multiplant operation (Scherer, 1980). The sources of these economies, according to Markusen (1984), are often found in firm-specific activities — including, R and D, advertising, marketing and management services — as opposed to plant-specific activities. A key characteristic of these different types of informational activities is their 'public good' or 'jointness' property as production inputs, whose benefits can be distributed across a large number of plants at little extra cost. These multiplant economies, Caves (1982) suggests, are the primary source of production flexibility in large multilocational firms. The 'jointness' aspect of informational activities enables firms to multiply geographical locations of direct production without reducing the marginal product of these intangible assets in existing plants.

The rationale for computer network innovations is, of course, to lower the average costs of producing and distributing the firm's informational inputs. As such, we would expect 'networking' to enhance production flexibility by increasing the level of output where average costs are least (Stigler, 1939); while an inverse relationship may still exist between firm size and flexibility, as Mills (1984) suggests, it is extremely plausible that binding profitibility constraints on the firm's expansion will operate at higher output levels. The corollary of this theoretical argument, for spatial analysis of the firm, is that computer networking will enable large multilocational firms to diversify geographically, and otherwise increase the geographical range of their production operations and markets.

It should be borne in mind that our discussion of economies of scale and scope, as they apply to multiproduct or multiplant firms, has focused purely on the costs of production and distribution as opposed to transactions costs. That is to say, we have remained faithful to the neo-classical analysis of economies of scope in a way that is consistent with Porat's (1978) 'four-factor' production function. The production flexibility argument, as such, derives from a technology-based theory of the multilocational firm. Economies of scope do not imply a need for production within a single firm (Teece, 1980), in the same way that technological interdependence favours vertical co-ordination of production, but not necessarily vertical integration. Thus, in discussing potential changes in the geography of the firm, we have not ventured an explanation as to why firms should multiply their geographical locations, or why new production operations should be, or not be, in common ownership (the emphasis of the transaction-cost approach). These questions were, however, not the centrepoint of this chapter, and they are addressed later in the book.

Conclusion

We know that telecommunications networks, like transportation systems, have a crucial influence on the geography of economic development owing to their space- and time-adjustment effects on productive, distributive and managerial operations. In different historical periods, telex and the telephone have played a *critical* enabling role in providing for production and locational flexibility in firms *and* government, without which the mass production economy could not possibly have evolved as it has done. In other words, telecommunications innovations have exerted a primary, rather than secondary, influence on the geography of capitalist economies.

Today, we are entering a new 'age of flexibility', and the geography of economies will once again be significantly influenced by technological advances in telecommunications. In this 'new age', we must reckon with the 'synergy' arising from technological convergence in telecommunications *and* computers, and this entails turning our attention to computer networks. In going beyond the telephone to computer networks, we encounter information flows which embody the inter-regional or inter-urban transfer of not just labour services (voice communications) but *capital* services (data communications) within, and between, firms and governments. It is this basic change — affecting the mobility of capital, rather than labour — which marks off computer networks from previous telecommunications technologies.

In the next chapter, I will consider what this 'hyper-mobility' of information capital signifies for some conventional approaches to modelling the spatial dynamics of economies. It will emerge from this theoretical critique that the spatial behaviour of information capital — organised into computer networks — is sufficiently problematic to force a re-examination of these conventional models, whose diagnostic and prescriptive 'outputs' have for long dominated regional policy thinking.

Notes

1. See the *Economist*, *Survey of Telecommunications*, 17 October 1987.
2. H. Godo, B. Netland and K. Oroy, *Storbrukere av Data Kommunikasjon i Norge*, TF-Report no. 50/87, Teledirektoratets Forskningsavdeling, Norway, 1987.

Chapter Four

Information Capital and Regional Economic Models[1]

Introduction

The chief object of this chapter is to kindle interest in the fundamental problems which the spatial division of information capital presents for regional economic models. What I will, however, seek to avoid is a closed system of generalisations about these 'problems', because this may convey the mistaken impression that I have ready-made solutions to them. The chapter should be seen simply as an initial attempt to identify the more obvious problems posed by the growing use of 'information technology as spatial systems', and the key considerations which need to be brought forward in future research on the regional dynamics of economies.

What is at issue here are the unstated, or 'implicit', assumptions which economic geographers (and 'spatial' economists) make about the location and use of fixed capital. These implicit assumptions in general, as Bliss (1987) notes, are where the 'vision' of researchers is typically located in their approach to economic problems. In quoting Schumpeter's conception of 'vision', Bliss states:

> Obviously, in order to be able to posit to ourselves any problems at all, we should first have to visualise a distinct set of coherent phenomena as a worthwhile object of our analytical effort. In this book, this pre-analytic cognitive act is called Vision' [Bliss, 1987: 295]

Of particular interest here is the geographer's or regional economist's 'vision' of capital. It is argued that traditional regional economic models implicitly assume that the use of capital is physically constrained by its location in geographical space. This is revealed, in particular, by what types of data are collected and used in empirical work — statistics on the physical stock of regional capital — and the models' policy prescriptions — initiatives for affecting a physical 're-balancing' of capital goods distribution in an inter-regional context. Indeed, my earlier allusions to the defectiveness of these models, where the information economy was concerned, were predicated on the obvious fact that 'networked' information capital does not obey these 'geographical laws' of capital location and use (Hepworth, 1986a, 1987a).

In another paper, promulgating this point about how information capital defies the latent 'geographical tyranny' of regional economic theory (Hepworth and Waterson, 1988), I was reminded by an anonymous referee that the 'target' of my erstwhile criticism — the specification of capital inputs in neo-classical growth, Keynesian income-expenditure and input–output models — was more elusive. The referee in question pointed out that for these theories 'to work', capital inputs need to be identified only in terms of regional *access* to the service streams generated by capital goods. This same point is made by Lachmann (1978: 58), as follows:

> There is, however, another equally important reason why we cannot accept a definition of capital structure in terms of the constant composition of capital combinations. Not the individual capital goods, but the service streams to which they give rise are the primary objects of our desires, and hence the ultimate determinants of the economic system. Capital goods are merely the *nodal points* of the flows of input (of labour and other capital services) which they absorb, and of output (of intermediate or final products) which they emanate.

The importance of this good–service dichotomy of 'networked' information capital for spatial analysis of technological diffusion was noted in Chapter Three. It is clearly not enough to define the inter-regional capital structure of the information economy merely in terms of the constant composition of capital combinations. The capital service streams, flowing as coded information over trans-regional computer networks operated by large multilocational organisations, have to be 'added in' to our calculations of productive capacity.

It now remains to elaborate these ideas step-by-step, beginning with the 'traditional' regional economic models alluded to above and their implicit assumptions pertaining to the location and use of capital. This is followed by a second and brief visit to Porat's (1977, 1978) 'four-factor' model of the production process, where information capital was formally introduced. It is shown that, although information capital may be a useful generic concept, Porat's reformulation of the production function can be criticised on the same grounds which apply to traditional 'two-factor' models. The centrepiece of my theoretical argumentation is the concept of 'the communicability of information capital', which complements rather than substitutes for our existing notions of factor mobility in geographical space. This 'communicability' characteristic, it is argued later, matters for geographical research wherever capital enters into aggregate analysis — for example, in the analysis of regional investment, or local labour markets, or changing urban form.

Conventional models of regional economies

Several frameworks have been used to analyse regional growth and development. These include the Keynesian income–expenditure approach, regional input–output analysis, neo-classical models of economic growth and analyses of inter-regional trade and migration. What is intended here is not a complete review but rather a discussion which is sufficient to appreciate the points made in later sections.

Both the Keynesian income–expenditure approach and regional input–output analysis are led by demand-side factors. The income–expenditure model starts with the accounting identity:

$$Y = C + I + G + X - M \tag{1}$$

where Y is regional income, C is regional consumption expenditure, I is regional investment expenditure, G is government expenditure in the region, X is regional exports and M regional imports. Provision may be made for income taxes, so that disposable income, $Y_d = Y(1-t)$ where t is the tax rate, constant for simplicity. It is also normal to have a linear consumption function, say $C = a + cY_d$, and a linear import function $M = b + m\,Y_d$, where c and m are the marginal propensities to consume and to import, respectively. By contrast I, G and X are all treated (in simple versions) as exogenously determined.

The focus of interest in the Keynesian-type model is on the regional multiplier, that is the effect upon income of an increase in autonomous expenditure of one sort or another. For example, a government policy change might dictate an increase in expenditure in the region. In the case of our simple model, the regional multiplier, k is given by:

$$k = 1/[1-(1-t)(c-m)] \qquad (2)$$

Hence, as Armstrong and Taylor (1985: 12) say: 'The critical variable in the regional multiplier formula is the marginal propensity to consume locally-produced goods (i.e. $c-m$).' The higher this propensity, the higher is the multiplier, and so the larger is the spillover effect of any given expenditure.

Before passing on, two features of this approach are worthy of our attention. The first is the important conventional distinction between consumption expenditure and investment expenditure, arising out of the (regional) accounting distinction between current consumption and additions to the capital stock, or investment. Notice that very different behavioural assumptions are made regarding these two expenditures, and these distinctions are maintained in more complex models. The second point is, to reiterate, that this model is very much concerned with the demand side — increases in demand for the region's output.

Although the focus of regional input–output analysis is also, conventionally, on the influence of demand factors, the approach and the insights offered differ greatly. Input–output analysis is very much concerned with the internal workings of the (regional) economy in the sense of inter- and intra-industry requirements for inputs. The basic equation is the following:

$$X_i = \sum_i X_{ij} + Y_I \qquad i = 1,2...,n \qquad (3)$$

which expresses total output of industry i as being the sum of its outputs sold to all industries j (including i) and to final demand Y_i. We can express this as:

$$X_i = \sum_i a_{ij} X_j + Y_i \qquad i = 1,2...,n \qquad (4)$$

where $X_{ij} = a_{ij} X_j$, q_j being the input–output coefficients $0 \leqslant a_{ij} \leqslant 1$. For example, if, in the region, industry j (motor car assembly) buys one-quarter of the output of industry i (battery manufacture), then $a_{ij} = 0.25$.

Equation (4) could be thought of as a purely descriptive model of the economy, true at any point in time as an identity — each firm's output goes to other firms in various proportions or to final demand. However, it is commonly used not as a description but rather as a predictive device, through the simple expedient of assuming constancy in the a_{ij}s. Then, writing it in matrix form:

$$X = AX + Y \qquad (5)$$

where $X = \{X_1 X_2 \dots X_n\} \quad A = \begin{bmatrix} a_{11} \dots a_{1n} \\ a_{n1} \dots a_{nn} \end{bmatrix} \quad Y = Y_1 Y_2 \dots Y_n$

(and braces denote a column sector).

Assuming the relevant inverse exists, from this we may write:

$$X = (I-A)^{-1}Y \qquad (6)$$

where I is the identity matrix.

The application of regional input–output analysis is now clear. An injection of a given amount of final demand in some sectors feeds through to others and so has implications for gross demand. In a way this can be thought of as akin to the Keynesian multiplier effects, but instead of looking at an aggregate impact, we may examine the effect on individual sectors.

Naturally, there are some implicit assumptions underlying this approach. Constancy of the a_{ij} implies constant returns to scale across all sectors (doubling Y doubles X). It also implies constant factor proportions in each industry and the absence of supply constraints. In connection with this, real-life phenomena such as multiregion, multi-product firms trading in intermediate goods cause a plethora of problems for the simple input–output approach. Generally, the openness of regions makes for difficulties.

Of course, the approach may be made more sophisticated so as to relax some of these constraints. For example, in modelling energy demand, Hudson and Jorgenson (1974) recognise the impact of relative price changes upon the nature of supply. Hence, they allow for non-constant input–output coefficients a_{ij} in a matrix like A, by estimating translog production models for each of their sectors. More directly connected to regional analysis are the mixed models of energy-economic-pollution interactions (see, for a review, Solomon, 1985).

Moving on to neo-classical regional growth models, these again may be traced back to origins in national economy models. Following Solow (1957) we start with an aggregate production function:

$$Y = A(t)F(K,L) \tag{7}$$

which incorporates neutral disembodied technical change, $A(t)$, and has capital K and labour L as aggregate facrtors of production. Logarithmic differentiation and a little re-arrangement yields:

$$\frac{d \ln Y}{dt} = \frac{d \ln A}{dt} + W_k \frac{d \ln K}{dt} + W_l \frac{d \ln L}{dt} \tag{8}$$

where W_k and W_l are capital's and labour's share

$$\frac{\partial Y}{\partial K} \cdot \frac{K}{Y} \quad \text{and} \quad \frac{\partial Y}{\partial L} \cdot \frac{L}{Y}$$

respectively. Assuming constant returns to scale means $W_l = 1 - W_k$. Then, if we define $y = Y/L$ and $k = K/L$ we may write equation (7) in intensive magnitudes as:

$$y_g = A_g + W_k k_g \tag{9}$$

where $x_g = d\ln x/dt$. The meaning of this equation is that the growth in output per head is related to technical progress and the growth in capital per head. As Armstrong and Taylor put it:

> Regional differences in the growth of output per worker are therefore explained by regional differences in technical progress and by regional differences in the growth of the capital/labour ratio [p. 56]

Obviously, this model concentrates on supply-side influences, both increased capital intensity and 'manna from heaven' technical progress, with the increase in output by assumption finding a ready market. As the derivation has shown, it also incorporates several rather restrictive assumptions.

All three models outlined above concentrate modelling effort upon the region itself. Obviously, they all make simplifying assumptions. More particularly for our present purposes, although these models do not make explicit assumptions about capital fixity, their operationalisations often do. Thus, it could be argued that production function, Keynesian, and input–output models assume of necessity that only access to capital services is required, wherever these services emanate from. However, the extension in practical work is then often made that the capital is spatially fixed within the region, used at that location, and used up in the course of production.

These assumptions are most explicit, perhaps, in the production function model, where regional differences in capital–labour ratios are seen as sources of technical progress. By assumption, if the regional capital–labour ratio is constant, growth in regional output per head comes about solely through exogenous technical progress. However, the assumption is also fairly clear in the input–output framework, where 'investment' is commonly treated as a final demand sector in the region. Imports into the region are allowed, but the standard fixed coefficients assumption imposes a complete lack of substitution between domestic and imported inputs. In the Keynesian models also, investment is a domestic element, though exports are treated equally as an injection with investment (in simple models). Imports usually depend directly on disposable income, and thus on consumer rather than producer demand.

The assumption that physical capital is fixed in location and used at that location also underlies data collation exercises such as Gleed and Rees (1979). They derive a collection of regional investment series by industry and, from these, a set of estimates of regional capital stock figures. It is envisaged that such estimates: 'will enable more extensive analysis of the effects of regional policy on the diversion of investment', and 'It is also hoped that the estimates of the regional capital stock will be of value in conducting technical efficiency comparisons between regions' (p. 330).

Yet, at the same time, most observers recognise that whilst some national economies are relatively closed, regional economies are almost necessarily open to trade. Thus, models of the type we have been considering do not tell the whole story:

A region must be treated as an open economy, and accounts constructed in a way that facilitates tracing the impact of external forces on local production flows. [Richardson, 1969: 232]

In particular, for our purposes, factors of production are likely to have some mobility.

However, there is rather an asymmetry in the literature regarding inter-regional migration. As Armstrong and Taylor (1985: 101) admit:

Our discussion concentrates almost entirely on labour migration in this chapter primarily because so little empirical work has yet been carried out on interregional movements of capital ... work on captial migration has been largely restricted to an investigation of the movement of manufacturing establishments between UK regions.

Why this asymmetry? It appears that the literature maintains the assumption of immobile physical capital. As Hoover (1975: 234) puts it:

Perhaps the most important difference between the processes of capital and labour migration lies in the fact that most capital has to be 'sunk' or invested in durable forms such as site improvements, building and production equipment before becoming useful. This major portion of the capital stock has virtually no capital mobility.

Gertler (1984) reinforces the view that this assumption is widely held (whilst he himself disagrees with it).

Finally, it is clearly implicit in most uses of the above models that capital is consumed in the course of production. Capital is assumed to depreciate so that there is a difference between net investment and gross investment, and flows of capital input are required to sustain production.

The 'information economy' model

Let us now return to Porat's 'adjustments' to the neo-classical production function which could be incorporated into Solow-type models of regional growth (see equation 7 for comparison). The following basic model of the (regional) production process, in which capital and labour inputs are differentiated by their information and non-information components, was introduced in Chapter Three:

$$Y = (K_I, K_N, L_I, L_N) \tag{10}$$

where Y = output
K_I = information capital inputs (e.g. computers)
K_N = non-information capital inputs (e.g. pick-up trucks)
L_I = information labour inputs (e.g. clerks)
L_N = non-information labour inputs (e.g. welders)

At this stage, we could expand (10) in the same manner as we expanded (7) to obtain (8) by logarithmic differentiation. Here, this would yield:

$$\frac{d\ln Y}{dt} = W_{ki}\frac{d\ln K_I}{dt} + W_{kn}\frac{d\ln K_N}{dt} + W_{li}\frac{d\ln L_I}{dt} + W_{ln}\frac{d\ln L_N}{dt} \tag{11}$$

This expresses growth in output as the result of growth in the various inputs. More revealing, however, suggests Porat, is to consider an explicit functional form for (10), namely the translog form. This is a second-order approximation to any production function, and it will involve terms in $K_I.L_I$, $K_I.L_N$, $K_N.L_I$ and $K_N.L_N$. Porat envisages that the coefficients on these interaction terms (which influence elasticities of substitution between factors) will provide interesting insights into the effect of information labour and capital inputs on marginal productivities and variations in effect across industries.

The assertion that information inputs are technically distinct from non-information inputs is central to Porat's more widely applied model of the information economy, which uses a standard national accounting approach and published statistics (Porat, 1977). In this variant of the Leontief-type input–output model, the information sector is 'partitioned' off from the rest of the economy for analytical purposes, and, consists of a primary and a secondary component — these being transactions in marketed information goods and services and in in-house information services produced by non-information firms, respectively. Here, values for the primary information sector are readily obtained from final-demand totals for selected information industries. However, the concept of information capital is explicitly introduced because of its particularly crucial role in devising indirect estimates for the value and structure of the secondary information sector. A value-added approach is adopted by which the size of the secondary information sector is calculated from two basic inputs, namely: the employee compensation of information workers in non-information industries together with depreciation taken on information capital in non-information industries.

The final expression of Porat's information economy model reveals its fundamental conceptual roots in Keynesian income–expenditure approaches (see equation 1). The economic impacts of new information technologies are examined using conventional input–output analysis (see equation 5). Schematically, it is represented as:

$$
\begin{bmatrix} X_N \\ X_P \\ X_S \end{bmatrix} = \begin{bmatrix} a_{NN}\,a_{NP}\,a_{NS} \\ a_{PN}\,a_{PP}\,a_{PS} \\ a_{SN}\,a_{SP}\,a_{SS} \end{bmatrix} \begin{bmatrix} X_N \\ X_P \\ X_S \end{bmatrix} + \begin{bmatrix} Y_N \\ Y_P \\ Y_S \end{bmatrix} \tag{12}
$$
$$
[V_N \ V_P \ V_S]
$$

where subscripts N, P and S refer to non-information, primary information and secondary information sectors, respectively; X and Y are gross output and final demand as before; V is value added; and the a_{ij}s are technical coefficients of which a_{PN}, a_{NS} and a_{SP} are believed to be zero (see Karunaratne, 1986).

Here, innovations in the new information technologies are treated as a component of final demand — gross fixed capital formation — with the usual column vector being split into a rectangular capital flow matrix. In Porat's original work on the US economy, disaggregated data on inter-industry flows of information and non-information capital goods were provided by the Bureau of Economic Analysis; similar inter-industry transactions data are used to estimate the model by Warskett (1981) and Karunaratne (1986) for Canada and Australia, respectively.

In sum, new information technology and its economic impacts are modelled by Porat like any other type of physical capital, in spite of his dichotomisation of factor inputs and sectoral partitioning of the economy. We are presented, therefore, with variants of the conventional economic models outlined earlier.

Given the previous extended application of these models to regional economies, the next logical step is to evaluate these revisions in a spatially disaggregated context. We examine, in particular, whether the basic assumptions about the location and use of information capital hold up if we were to analyse the regional dynamics of the information economy as it has been formulated.

The spatial communicability of information capital

Information capital, as it is defined by Porat (1977) for accounting purposes, includes a long list of physical assets which holds no interest for us — for example, office buildings and furniture, X-ray equipment and even clocks. We are restricting our interest to the new information technology as a primary form of information capital used by large multilocational organisations. Furthermore, of immediate concern are the implications of the spatial division of information capital in computer networks for modelling the regional dynamics of economies.

We should note, at the outset, that private computer networks are already an established component of the capital stock. For example, Canadian evidence indicates that these types of networks emerged in the early 1970s, exhibited accelerated rates of adoption in the second half of the decade and diffused less rapidly in the 1980s (Table 4.1).

Table 4.1 *The adoption of computer network innovations in Canada*

INDUSTRY SECTOR	PERIOD NETWORK INSTALLED (No. Firms)		
	Pre-1975	1975–80	Post-1980
1. Primary Industries	2	3	3
2. Manufacturing	2	7	5
3. Transport, Communications, Other Utilities	2	0	3
4. Trade	1	2	1
5. Finance, Insurance, Real Estate	0	2	1
6. Computer Services	1	1	0
7. Other Services	0	1	0
ALL SECTORS – Number	8	16	13
– % Period	(22)	(43)	(35)

Source: Hepworth (1987a). Data are from a questionnaire survey of large Canadian firms carried out in 1984.

Further findings from the same source (Hepworth, 1987a) confirmed that network innovations are used by large firms in all sectors of the Canadian economy, including manufacturing industry; network applications were also found to be progressing 'upwards' through the corporate informational hierarchy, from routine functions (e.g. order entry) to strategic functions (e.g. decision-support for senior corporate management) (Table 4.2).

There is no reason to believe that the diffusion path of computer network innovations has been significantly different in other industrialised countries. As such, these innovations are already real elements of the regional capital stock, as opposed to future or futuristic considerations which can be temporarily set aside in our calculations of productive capacity and its attendant geographical distribution. Indeed, it is these very 'calculations', providing the base data for empirical applications of the conventional regional economic models outlined earlier, which are made extremely problematic by what I call the 'spatial communicability' of information capital.

To retrace our steps, we saw that computer networks are characterised by a spatial division of information capital. This capital structure, technically articulated in the 'architecture' of computer networks, derives from the economies of scale and scope in information-handling which 'networking' makes available to organisational users. Computer networks, as noted earlier, are basically resource-sharing systems which span locations and functions. Their spatial incidence is then indicated by the 'network topology(ies)' implemented by a particular firm (or government agency).

Table 4.2 *Application trends of computer network innovations in Canada*

ORGANISATION FUNCTIONS (by application)	CURRENT APPLICATIONS USAGE (No. Firms)			GROWTH AREAS OF APPLICATIONS USAGE (No. Firms)		
	Yes	Indefinite	No	First	Second	Third
1) Strategic Management	25	9	9	13	21	6
2) Routine Operations	37	2	4	8	8	23
3) Administrative Support	17	14	12	15	11	12

Source: Hepworth (1987a). Data are from questionnaire survey of large Canadian firms carried out in 1984.

PETROLEUM COMPANY
Network applications: market research, customer services, inventory control accounts, message services, and refinery simulation models

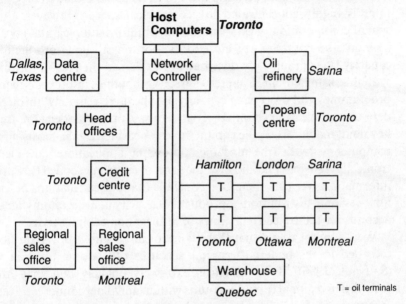

T = oil terminals

FOOD PROCESSING COMPANY
Network applications: financial control, business planning, customer services, inventory control, accounts, payroll, order entry

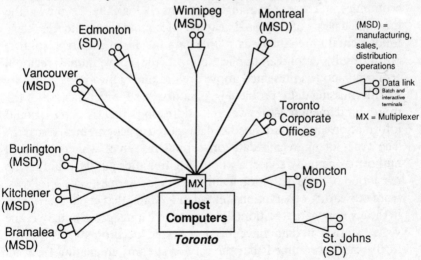

(MSD) = manufacturing, sales, distribution operations

Data link
Batch and interactive terminals

MX = Multiplexer

Figure 4.1 *Real network topologies showing information capital concentration*
Source: Hepworth (1987a)

Now, even from a cursory look at the topologies of real computer networks, such as the Canadian examples presented in Figure 4.1, it is glaringly obvious that physical information capital in large corporations is highly concentrated in geographical space. However, this spatial configuration of physical capital, or information capital *goods* (computers, multiplexers, etc.), does not reveal how productive capacity is geographically distributed. I explained the reason for this spatial 'mismatch' in Chapter Three: computing resources (data, programmes and processing capacity) can be used 'remotely' through data communication channels. That is to say, and to underscore this key point made earlier, the capital *services* provided by the 'hosts' in a computer network (the machines processing applications) are electronically transmissible and take the form of electronic, coded flows of information. We can, therefore, conclude that our calculations of the inter-regional distribution of productive capacity in the economy must account for these electronic flows of information capital services.

We know for a fact that these types of information flows are not recorded in the (external) capital accounts of national economies (Sauvant, 1986). It is, therefore, extremely unlikely that they are recorded in regional capital accounts within individual countries, given that these sources of disaggregated data are reputed to be poor. In other words, for the present at least, we have no statistical data on the total information capital used at the level of national or regional economies. These 'missing' statistics logically signify the incompleteness of regional capital accounts and, hence, their fundamental unreliability as source data for use in empirical studies. As a result, empirical applications of the conventional regional economic models presented above are ultimately thwarted.

This 'missing data' problem is, of course, a superficial aspect of the more fundamental issue we are addressing, that is: how should information capital be defined and measured in geographical research? The types of aggregate data on capital resources which statistical authorities actually collect and publish are, after all, pursuant to the resolution of these definitional and measurement questions. It follows, from our earlier observations on 'missing' data, that capital resources in the economy are traditionally equated to capital *goods* and, over the years, statistical data have been collected accordingly. We have, furthermore, presumed that the service streams emanating from all forms of capital goods follow a geographical 'short circuit' — that is, they specifically occur and take direct effect in the micro locale where a given piece of equipment is physically implanted. Thus, our past calculations of the geography of productive capacity have been predicated on the assumption that the location of capacity (capital goods and services) and the location of capital goods are effectively one and the same.

We can discern this implicit assumption, pertaining to the spatial 'short circuitry' of capital, in empirical studies of the geography of capital formation and its differential effects on regional economic growth. A recent Canadian study by Gertler (1986b), for example, identifies the inter-regional geography of capital merely in terms of the spatial distribution of fixed investment (capital goods). Capital mobility is, similarly and routinely, analysed in terms of the physical relocation of whole firms or parts of firms, based on establishment, closure and plant migration data (see, for example, Bluestone, 1987). In other words, the spatial dynamics of information capital — as I have conceived of it here, in terms of the inter-regional flow of capital services mediated by computer networks — appear to have been overlooked in macroeconomic models of regional development.

These inter-regional flows of captial services in computer networks increasingly occur in 'real time'. As stated earlier, they take the form of electronic communications, so that I prefer to use the term 'communicability' when referring to the movement characteristics of 'networked' information capital. This term, in my view, captures the essential quality of the spatial dynamics of information technology in computer network form. It also serves to highlight its qualitative difference from alternative conceptions of capital mobility, which are more appropriate to analysing the physical relocation of firms or fixed capital investment.

Thus, we are confronted by a powerful new factor in regional economic development, namely, the spatial communicability of information capital in computer networks. The economic importance of this factor, and thereby the need to take stock of its role in regional development, will grow enormously as computer networks become increasingly central to management, production and distribution functions in firms across all sectors (Cohen and Zysman, 1987). From the foregoing discussion, it is clear that the new and different spatial attributes of 'networked' information capital pose important questions for regional analysis; in particular, researchers will need to look closely at the implicit assumptions which underpin regional economic models as they relate to the location-use characteristics of capital resources.

'Space of flows'

We have, in essence, two associated geographies of information capital in computer networks. The first is a geography of capital goods *fixed* in physical space, which is familiar to all of us. The second is a lesser known geography of capital services *flowing* in information space. Elsewhere, I have referred to this second geography as 'electronic space' (Robins and Hepworth, 1988a). Its ascendance as an economic field has already been noted by some other geographers (see, for example, Webber, 1980; Nicol, 1985).

The spatial communicability of information capital is, indeed, alluded to by Friedman and Wolff (1982) in discussing the differential mobility of factors of production: 'electronic communications may be a substitute for physical relocation' (p. 317). We also have Castells's (1985) imagery of the 'new space of production and management' created out of computer network innovations, as follows:

> We still have to deal with space, but we increasingly observe a space of flows substituting a space of places. Thus, a hierarchy of functions and power positions structures the territory across the nation and across the world, separating functions and units of production, distribution, and management to locate each one in the most favourable area, yet articulating all activities through a communications network. [pp. 14–15]

The 'space of flows' articulated by computer networks is, of course, the space of all of those activities which bear the 'tele-' label — for example, tele-shopping, tele-banking, tele-conferencing, tele-working and so on. These types of 'tele' activities are enabled basically by computer network applications. In addition, there is the 'on-line' industry itself (Dordick, Nanus and Bradley, 1981) and the complex systems of electronic funds transfer which support the world's capital markets (Langdale, 1985). The so-called 'just-in-time' (order entry) systems in manufacturing, which have attracted the interest of industrial geographers (Estall, 1985), are also computer network applications. Indeed, the 'automated factory' (Freeman, 1988) and

'flexible production' methods in general (Camagni, 1988) are technologically based on the combined application of local area (site-specific)
and wide area (inter-city) computer networks in manufacturing firms.
In other words, the geography of information space articulated by
computer networks has significant horizontal (sector-wide) and
vertical (function-deep) dimensions in the economy.

We clearly need to chart out this new 'electronic' geography of
capital formation. This is essential to different areas of geographical
analysis, as underlined by Gertler (1986b: 524):

> Given the apparent importance of capital formation in determining the
> interregional geography of productivity, migration, prosperity and equity,
> it would seem that a sound understanding of the determinants of local or
> regional investment rates is of crucial significance to policy makers.

In this chapter, and in Chapter Three, I have provided a technical
description of 'information technology as spatial systems', which may
prove useful to future research on the geography of capital formation.
Although my heuristic discussion of the spatial behaviour of 'networked' information capital focused on some general models of the
regional economy, its main points are obviously relevant to capital-
modelling in partial analyses of the spatial dynamics of economies.

By way of illustration, Bell and Hart's (1980) model of the regional
demand for labour services in UK manufacturing has net capital
formation (along with manufacturing output and past employment
levels) as a key dependent variable in the short-term employment
equation. As argued before, the values assigned to these regional
capital inputs will, in future, need to account for the inter-regional
flows of information capital services in computer networks operated
by manufacturing firms. This 'adjustment' is most obviously necessary
for countries where these types of econometric models are routinely
used to evaluate the efficacy of capital subsidies or investment grants,
as elements of regional policy designed to reduce spatial disparities in
employment opportunities.

The above comments suggest, as well, that the distribution of capital
resources in the space economy depends not only on the transportation
infrastructure — for the physical movement of capital goods — but
also on the advanced telecommunications infrastructure — for the
'electronic' movement of attendant capital services. As such, where the

provision of capital resources is a significant element of regional policy, infrastructure planning must include 'blueprints' for both telecommunications and transportation. In other words, our computer network model of information capital implies that there can be no 'trade-off' between these infrastructures, where they are used as regional policy instruments. They are complementary systems of capital distribution in what Castells (1985) calls 'the new space of production and management'.

The applications of regional econometric models are, of course, extremely wide-ranging in terms of the social, political and economic issues they address. The following list of 'important questions' is, for example, addressed by Clark, Gertler and Whiteman (1986: 311) for the US:

> First, how sensitive is the regional rate of growth of capital to the cost of major inputs such as labour? Secondly, what are the major determinants of capital intensity in regional production? Thirdly, what are the interrelationships between regional growth, technological change and the distribution of the product between workers and the owners of capital?

All of these questions are 'answered' by Clark *et al.* on the conventional basis of multiple regression analysis. Importantly, each and every equation in their series of regressions contains a capital stock variable measured as the dollar value of fixed investment (or net additions thereof). Again, with the diffusion of computer network innovations in US manufacturing industry (see, for example, Cohen and Zysman, 1987; Ayres, 1984), the future usefulness of these partial models of regional economic development will depend on whether adequate account is taken of interstate flows of information capital services.

The same order of attention to the new and different spatial behaviour of information capital will be needed in modelling the 'impact' of information technology on urban growth and form (Brotchie, Hall, Newton and Nijkamp, 1985). Here, the communicability of information capital presents firms with further opportunities for 'trading-off' transportation in favour of telecommunications, subject to the necessity of maintaining face-to-face contacts for 'non-routine', information-processing work functions. In Greater London, for example, 'telecommuting' is now increasingly considered as a possible panacea to the national capital's syndrome of 'overheated' markets for labour and land/property and transport congestion (Hall, 1989). Telecommuting projects in Los Angeles have a similar rationale — that is, to offset the agglomeration diseconomies of scale associated with urban growth, by reordering 'travel- to-work' areas (the local labour market) in terms of electronic rather than physical space.[2]

Along these lines, Kutay (1986a) has attempted to show that the optimum location of offices in information economies is affected by the distance-adjustment potential of advanced telecommunications. The stated intent of Kutay's analysis is to demonstrate:

> how the office modifies its 'telecommuting' distance with employee residences and 'telecommunications' distance with multi-centres as the volume of information production expands. [p. 560]

Notably, her 'information production function' is of the conventional Cobb–Douglas type, with office labour and office capital other than telecommunications technology being the sole factor inputs. We must, therefore, assume that office capital here includes computers and terminals, and perhaps communications-processing equipment — front end processors, concentrators, multiplexers and so on.

Now, as we have seen, these investment resources in the office are subject to the economics of 'networked' information capital. As such, in empirical applications of Kutay's model, our calculations of office productive capacity would need to include information capital services (computing resources) delivered or accessed 'remotely' over the firm's computer network. If these service streams are not accounted for, the observed behaviour of Kutay's information production function over a given range of output, and thereby the locational model's predictions, would clearly be spurious.

Conclusion

In deference to the anonymous journal referee mentioned in this chapter's introduction, we could simply accept the argument that *in theory* the 'conventional' models targeted for our critique can accommodate the spatial communicability of information capital. These models, so the referee's argument goes, require only that full account be taken of the capital resources (goods and services) used in a given region for empirical applications. Indeed, the same could be said for all of the partial models of urban and regional development, which I have referenced in order to illustrate the potential extent of the 'missing capital' problem.

What this 'problem' reveals, however, is the urgent need to change our 'vision' of capital and its spatial behaviour. In this regard, 'bandaid' repairs to conventional regional economic models act only to 'brush' the profound issues raised by information capital 'under the carpet'. The adumbration of so-called 'information production functions', similarly tends to distract from the conceptual and analytical problems posed by the indivisibility characteristics of information capital, which render marginal theoretic approaches inadequate (Stiglitz, 1985).

In essence, we are witnessing the emergence of two geographies of capital formation and mobility which are inextricably linked, as information technology is incorporated into the capital structure of firms. This aspect of spatial differentiation, as opposed to technological differentiation in regional capital stocks arising (say) from functional specialisation in multiplant firms, has not yet been studied by economic geographers. Clearly, if the spatial implications of information technology for production organisation in *all* sectors are to be identified, we need to develop a unified analytical framework for exploring not one but two geographies of capital.

In the next chapter, I will make a small invasion into this 'new space of production' through empirical description of computer networking in large multilocational firms. These case studies are, of course, inherently limited in terms of what they reveal about the total relationship between technological change and urban/regional development. They may, nevertheless, provide some useful evidence on the workings of the 'new' space economy, as an initial basis for defining future research agendas.

Notes

1. The first half of this chapter is reproduced from the published article 'Information Technology and the Spatial Dynamics of Capital', *Information Economics and Policy*, 3, (1988), pp. 143–63. I am very grateful

to Elsevier Science Publishers and the article's co-author, Professor M. Waterson of Reading University for allowing me to reproduce the relevant text.
2. Telecommuting: 'A Pilot Project Plan. Department of General Services, State of California'. Paper available from JALA Associates, Los Angeles, California, June 1985.

Chapter Five
Computer Networks in Multilocational Firms

Introduction

Large multilocational firms still dominate the corporate sector of advanced industrialised economies. Their usage of information capital — in the form of private computer networks — enables firms to achieve higher levels of production and locational flexibility (see Chapter Three). In this chapter, I examine how computer network innovations may affect the spatial and functional organisation of large multilocational firms, using an assortment of case study and anecdotal evidence.

The first section argues that computer network innovations will change the economics of multilocational firms by lowering the costs of producing and distributing information services. A major result of this technological change is to permit firms to increase their 'span of control' in geographical and market space, through more 'flexible' technical and social divisions of labour. Some case study evidence is then presented by way of illustration, and to illuminate these important changes in the geography of enterprise which are made possible by information technology. The last sections of the chapter elaborate on the results of this empirical description, focusing on the competitive strategies and organisational structure of the so-called 'network firm' (Antonelli, 1988b).

Information services and locational flexibility

It was suggested, in earlier chapters, that computer networks enable large multilocational firms to achieve significantly higher levels of locational flexibility, owing to the greater economies of scale and scope which the technology permits in the production and distribution of information-related services. The main sources of multiplant economies, as noted by Markusen (1984: 207), 'are often found in firm-specific as opposed to plant-specific activities. These firm-specific activities include things like R and D, advertising, marketing, distribution, and management services.' The efficiency advantage conferred on multilocational firms arises, then, from spreading the costs of these centrally produced intangibles over several geographically dispersed sites of production activity (Caves, 1982).

The general effect of computer network innovations is to lower the costs of producing and distributing information services within the multilocational firm (Rada, 1984). These enhanced economies of scale and scope, deriving from the sharing of information and specialised physical assets (computers and telecommunications facilities), provide the firm with opportunities for reducing the minimum efficient scale of branch operations (remote plants, sales offices, etc.) and extending their degree of geographical dispersal (Antonelli, 1988a; Waterson, 1987). As such, we can expect a higher level of locational flexibility in multilocational firms with the diffusion of computer network technology. This would apply to multilocational firms in all sectors, including manufacturing, but the most visible effects would clearly be found in information-intensive industries — for example, finance — and in the producer services sector more generally (Howells, 1988; Moss, 1987).

This production flexibility argument for the greater locational choice permitted by computer networking applies also to multinational corporations (MNCs) (De Meza and Van der Ploeg, 1987; Markusen, 1984). Here, the enhanced 'transportabiliy' of information services (Rada, 1984) and the absence of tariff restrictions on the international movement of informational inputs and outputs over corporate networks combine to increase the locational flexibility of MNCs in a global context. According to Sauvant (1986: 8), for example:

> Because of the nature of services, their production and consumption have normally to occur at the same time in the same place at that this represents a serious obstacle to trade. Real-time, interactive communication via transnational computer-communication systems changes this situation. By collapsing time and space (at decreasing costs) transactions can take place

at the same time in different places. As a result, the *tradeability* of certain services increases considerably affecting especially such key business services as banking, insurance, accounting, design and engineering, legal services, management consulting, and, of course, data services themselves.'

It is estimated that 80–90 per cent of so-called 'transborder data flows' is generated by intra-firm services transactions,[1] and manufacturing firms are known to be major users of transnational computer network technology (Espeli and Godo, 1986; Antonelli, 1981).

While computer network innovations may enable firms to increase the geographical scope and number of their productive operations and final markets, the increasing costs of organising transactions within the firm have the potential to act to limit its expansion. In this case, what Coase (1937) calls 'diminishing returns to management' posed by bounded rationality can be (partly) offset by the use of information technology. According to Coase:

> As more transactions are organised by an entrepreneur, it would appear that the transactions would tend to be either different in kind or in different places. This furnishes an additional reason why efficiency will tend to decrease as the firm gets larger. Inventions which tend to bring factors of production nearer together, by lessening spatial distribution, tend to increase the size of the firm. Changes like the telephone and the telegraph which tend to reduce the cost of organising spatially will tend to increase the size of the firm. All changes which improve managerial technique will tend to increase the size of the firm. [p. 343]

In so far as computer networks are superior control technologies to the telephone or the telegraph, they may act to increase the optimum size of firms and (probably) their geographical scope (Antonelli, 1988a). This increasing span of control effect on firm size is highlighted by Williamson (1986a), who states elsewhere that 'improvements in information technology commonly favour internalisation (of transactions, or a greater optimum degree of vertical integration) and more extensive internal auditing and controls — at least in the short run' (Williamson and Bhargava, 1986: 56). Similarly, Silver (1984) suggests that 'the introduction of modern computer networks would probably operate to [promote vertical integration] by providing more effective control over shirkers' (p. 47).

The powerful capabilities of computer network technology for enhancing management control over a more geographically distributed labour process in multilocational firms have been highlighted, for example, by Aglietta (1979). We would expect more complex, occupationally differentiated, spatial divisions of labour to evolve, as firms exploit the greater flexibility of network-mediated systems of work organisation. In this case, the enhanced capabilities of computer networks for metering and synchronising the performance of geographically dispersed work tasks may reduce the need for face-to-face contacts in certain occupations. And, as Schoenberger (1987), for example, suggests, network-mediated contact systems may substitute for inter-personal communication not only within firms but also between firms (linked by inter-organisational computer networks):

> We must, therefore, reckon with the possibilities that improvements in information and communication technologies may reduce the need for direct personal contact. This is most clearly the case for transmittal of changing product specifications within and across firms via, for example, linked computer networks. [p. 207]

The use of inter-organisational computer networks may, of course, reduce the optimum size of the firm, if the technology lowers the transaction costs associated with using the market more than it lowers the costs of organising transactions internally (Silver, 1984). However, given the transactional difficulties that attend market exchange in information (Teece, 1980), we would expect inter-organisational networking to lead to further refinements to the inter-firm or social division of labour, but with these tendencies being highly differentiated by function and spatially (Estrin, 1985, 1986). In the case of order-entry applications, such as 'just-in-time' systems in manufacturing, the frequent need for exchanging product-related information between contracting firms may encourage spatial agglomeration tendencies (Estall, 1985; Meyer, 1986). However, in the case of highly innovative industries or the emerging on-line information industries, the need for spatial proximity between contracting firms and the degree to which transactions are internalised will depend particularly on 'information impactedness' problems surrounding the development, manufacture and marketing of new products, and their unfamiliar methods of utilisation (Caves, 1982).

Indeed, rather than the simple spatial formula which 'just-in-time' network applications may appear to present, we should expect a more complex set of locational and organisational dynamics to arise out of computer network innovations. Consider, for example, Camagni and Rabellotti's (1988) findings of organisational change in the case of the clothing/textile industry in northern Italy:

> The setting-up of information [technology] networks within either large-scale vertically integrated companies, or systems areas, has considerably enhanced the potential for coordinating remotely located production phases in the textile cycle, as well as collecting and spreading out market-originated information ... The introduction of information technologies has largely affected the classic organisational alternative 'make or buy', which has confronted the textile/clothing industry at revamping time. With regard to organisation costs, more and speedily transferable information on-hand, allow a far easier control over the enterprise system and foster a functional integration. As concerns transaction costs, information technologies facilitate interactions both inside the industry, and outside the market. [p. 23]

In sum, what we expect to arise from computer network innovations is an extremely complex picture of organisational and spatial tendencies, rather than centralisation or decentralisation and vertical integration or disintegration, and so on. The case study evidence presented below should, as stated earlier, be treated as illustrative of particular developments, and it only partly helps to answer some of the important theoretical questions already raised here. We have, nevertheless, some provisional theoretical guidance to carry forward, and at least where the reader encounters an empirical vacuum in the following pages of discussion it will be readily discernible.

Case studies of technological innovation

As part of my doctoral research on the 'Geography of the Information Economy', at the University of Toronto, I carried out a number of case studies on computer network innovations in some of Canada's largest companies. This sample included: Labatt Breweries, International Business Machines (IBM) Canada, Bell-Northern Research and the *Globe and Mail* Newspaper (manufacturing); Imperial Oil Canada (resources-manufacturing); Air Canada (transportation); Sears Canada (retail trade); Canadian Imperial Bank of Commerce (CIBC) (financial services); and I.P. Sharp Associates (computer services). All research material was collected through 1984 from two-stage personal interviews with information technology (IT) specialists and other managers in these companies, correspondence and follow-up telephone interviews. I also gathered background material on each company from annual reports, market and industry publications and press clippings.

The interviews themselves were highly constrained by confidentiality agreements, however, they produced the following types of qualitative and quantitative information:

(a) the *organisational context* in which the technology is embedded, including the locus of management decision-making and the company's competitive environment;

(b) the *topology of the network* implemented by the company, focusing on the spatial and technical characteristics of the 'backbone' system;

(c) the *applications* of the network, including routine and strategic applications, such as payroll-processing and decision-support functions;

(d) the geographical distribution of the company's *informatics manpower*, including all higher and lower order personnel associated with operating the network to produce in-home computer and communications services — for example, systems analysts and data entry operators.[2]

The basic objective of these case studies, whose limitations are relatively obvious, was to develop some insights into the spatial and economic aspects of computer network innovations at the level of the multilocational firm. My intention was, then, to identify a few elements of a research agenda on technological change in the information economy, in terms of indicating some possible instruments of empirical analysis (surveys/case studies) and some basic questions which might merit closer attention in future investigations of the impact of information technology on urban and regional systems. A complete discussion of this case study research, and its relationship to my supporting questionnaire survey work on 'Computer Networks in

Canadian Firms', is to be found in the original doctoral dissertation (Hepworth, 1987a).

Given the space limitations of this chapter, I offer only a summary of these case study findings, partly reproducing some material from published articles (Hepworth, 1986a, 1989) and newly presenting other results. For convenience, I have tabulated this second set of case study findings (Table 5.1; see Appendix B for diagrams), which are used to illustrate and support various points brought out by my general discussion of concepts and factual observations. A detailed reporting of three case studies — the *Globe and Mail* newspaper, Bell-Northern Research and I.P. Sharp — is incorporated directly into the main text under separate headings. Before then, I have the following basic comments to make on the case study research, particularly with respect to its limitations.

Table 5.1 *Summary findings of computer networks in selected Canadian firms*

COMPANY DESCRIPTION	TYPE OF COMPUTER NETWORK	APPLICATIONS	REPORTED IMPACTS
SEARS CANADA Canadian subsidiary of Sears Roebuck, Chicago. Sector: Retail trade. Head office: Toronto. Employment: 55,000. Revenues: $3,300 m.	(i) Point-of-sale terminal net (Links stores directly to Toronto head offices; remote front-end processors only).	At sales outlet: processing customer transactions, credit authorisation and inventory management. At head office and sales analysis, truck-distribution scheduling, accounts and financial analysis, personnel, etc.	Lower operating costs in areas sensitive to interest rates; inventory and cash flow control. Lower (fuel) costs in goods distribution (transport schedules). Efficiency gains from functional integration and 'remote' control of distribution outlets.
	(ii) Inter-organisational net: retailer-supplier order entry links.	Purchase of telecommunications services from Canadian carriers for internal use. Purchase of merchandise from manufacturing suppliers for resale.	Lower costs of processing intermediate market transactions; lower purchase prices from manufacturing suppliers due to improved production scheduling; lower inventory costs.
	(iii) Inter-organisational net: parent-subsidiary links.	Financial reporting to parent company, Sears Roebuck. Transnational acquisition of computer software, from parent to subsidiary company.	Higher levels of parent-subsidiary control enabled by fiscal integration, reinforced by standardisation of software. International economies of scale in software development.
	(iv) 'Teleshopping' (Compushop): links to final customers.	Direct sales to 'home' customers: order entry, credit authorisation, direct debit payment and delivery scheduling.	Lower transaction costs in final markets; improved customer information and 'target marketing'; increase in Sears' credit card business.

COMPANY DESCRIPTION	TYPE OF COMPUTER NETWORK	APPLICATIONS	REPORTED IMPACTS
AIR CANADA Crown corporation. Sector: Air transport. Head office: Montreal. Employment: 21,500. Revenue: $2,570 m.	(i) Distributed net (links remote concentrators in sales offices, airport plants and Winnipeg financial department to Toronto and Montreal computer centres/ offices).	At Toronto: reservations, flight movement, pricing/ ticketing, seat selection, load planning and flight information (airport). At Montreal: flight planning aviation weather, human resources, material availability and all corporate functions (incl. remote on-line support of Winnipeg financial/accounts of offices).	Revenue expansion enabled by high-volume transaction processing in passenger reservations; increased competitiveness due to improved customer services. Lower costs of inventory (airplanes) and fuel permitted by more efficient flight planning. Efficiency gains from higher levels of integration and control in national and international operations.
	(ii) Inter-organisational net: travel industry links.	Resale of data base services (reservation system) to travel-related industries: hotels, car hire, road and rail transportation and travel agencies.	New and growing source of revenue expansion; vertical relations with travel industry essential to competitiveness in airline passenger market.
	(iii) Inter-organisational net: SITA (Société International de Télécommunications Aéronautique).	Resale of surplus 'mobile' inventory (cargo and passenger seats) between airlines; reservations for multi-airline flight pattern.	Lower inter-airline transaction costs and improved efficiency in 'mobile' inventory management, permitting internationalisation of market.
CANADIAN IMPERIAL BANK OF COMMERCE (CIBC) Chartered Bank. Sector: Banking. Head office: Toronto. Employment: 32,615. Revenues: $7,600 m.	(i) Integrated net: data centres in regional offices linked to main computer complex at Toronto headquarters; regional sub-nets support branch banks.	Cheque processing (data capture and central up-date of customer accounts). Visa card transactions. Instant or automated teller machines. Direct-debit payments (salary payments of client firms). Other banking services: mortgages, loans, etc. Internal applications: routine office support and decision-support for corporate management.	Lower transaction costs in cheque-processing. Diversification of services (commission revenues). Lower transactions costs; more transactions (market) through flexible access (location-hours). Improved competitiveness and lower transaction costs in business client sector. Diversification of services (revenue expansion). Higher levels of office productivity; enhanced management and control of diverse and extensive branch banking net.
	(ii) On-line banking net: links branch bank terminals to Toronto computer centre directly.	Interactive on-line support for branch bank tellers: customer account up-date/ withdrawals.	Lower transactions costs/ higher labour productivity in branch banking net; improved customer services and marketing of new products at local level/ increased competitiveness.
	(iii) Inter-organisational net: public data base	Purchase of financial and business data.	Enlargement of 'information space', permitting

COMPANY DESCRIPTION	TYPE OF COMPUTER NETWORK	APPLICATIONS	REPORTED IMPACTS
	services (I.P. Sharp, see case study in text).	Financial reporting by foreign subsidiaries and branch offices via service bureau net communications.	internationalisation. External economies of scale and scope in global transactions with foreign offices, permitting further internationalisation.
	(iv) Inter-organisational net: S.W.I.F.T. (Society for Inter-Bank Financial Transactions) and International VISA.	Inter-bank transactions: customer and bank transfers, foreign exchange, credit letters, etc. (SWIFT). Credit card transactions processing (VISA).	Horizontal integration creates scale economies in inter-bank transactions and lower costs, permitting internationalisation. Transnational scale economies in transaction processing.
	(v) Inter-organisational net: automated teller services (other banks).	Automated teller transactions processing.	External scale economies in transactions processing; interconnection with foreign bank nets permits internationalisation.
	(vi) Transnational data links with foreign subsidiaries and sales offices.	Monitoring transactions against customer and country lending limits; financial and investment reporting at global level.	Foreign market expansion and arbitrage opportunities in world capital markets; geographical diversification permits higher portfolio returns and reduced risk.
INTERNATIONAL BUSINESS MACHINES (IBM) Canadian subsidiary of IBM, Armonk (New York). Sector: Computer manufacture. Head office: Markham, Ontario. Employment: 11,500. Revenue: $2,100 m.	(i) Distributed net: computer centres at regional offices linked to multi-site computer complex at Markham headquarters. Note: IBM operates 3 other computer nets in Canada: for manufacturing, R & D and computer services. The case study net supports administrative and marketing operations only.	At regional sales/technical offices: business communications (electronic mail, on-line retrieval of customer and product information) plus 'decision-support' applications (eg. business graphics and planning). At Markham headquarters: both of the above application types plus 'operations' applications – eg. payroll, order entry, inventory control, planning and budget, customer accounts and histories, etc.	Enhanced competitiveness through direct customer contact in new and complex product areas: technical support in implementation and after-sales service. 'Captive' market for new product lines and other value added services. Higher levels of office productivity through distributed processing (faster response times in computing). Cost savings in inventory management. Economies of scale and scope in producing information services centrally: product and customer records and administrative functions, etc., permitting specialisation and less-bureaucratic regional offices.
	(ii) Transnational data links between Markham and US parent company's head offices.	Order entry for specialised components. Global electronic mail services. Parent-subsidiary financial and market reporting.	Lower inventory costs. Lower global communications costs. Higher levels of parent-subsidiary fiscal control and overall efficiency gains through vertical integration.

COMPANY DESCRIPTION	TYPE OF COMPUTER NETWORK	APPLICATIONS	REPORTED IMPACTS
LABATT BREWERIES Operating company of Labatt manufacturing-services conglomerate Sector: Beer manufacturing Head Office: London, Ontario Employment: 4,300 Revenues: $1,000 m	(i) Distributed net: remote minicomputers link plants to regional offices; regional offices linked to computer centre at London headquarters	Minicomputer in each region supports manufacturing plant and ancillary offices: production scheduling, inventory control, accounts payable/receivable, payroll, etc. London computer centre: administrative (word processing and electronic mail), business planning, financial control, market, research and advertising, and all corporate functions.	Lower inventory costs through more efficient production scheduling; lower administration costs in ancillary regional offices; enhanced market knowledge and sales forecasting. Higher labour productivity in management and administrative occupations; more effective market research and product development; enhanced control of regional operations; efficiency gains from functional integration at level of firm.
	(ii) Inter-organisational net: link to public data base vendor.	Daily, on-line purchase of political, economic, business and financial data.	More efficient monitoring of firm's environment: intelligence on market rivals (Molson and Carling), government policy, industry news, capital market conditions, etc.
	(iii) Transnational data links to new regional offices in Buffalo and Chicago.	Order entry, sales analysis and accounts receivable (plus routine office applications).	Market expansion in US permitted by lower transaction costs of sales offices; improved market intelligence on US product demand.
IMPERIAL OIL CANADA Canadian subsidiary of Exxon Corporation, New York. Imperial Oil's national subsidiaries: Esso Resources, Calgary; Esso Petroleum, Toronto; Esso Chemicals, Toronto; Sector: Oil and gas production. Head Office: Toronto. Employment: 14,700 Revenue: $8,615 m	(i) Corporate net based on Toronto and Calgary computer centres (Links head offices).	Parent-subsidiary reporting within Canada: production market and financial data consolidated at Imperial Oil head offices. Centralised applications: finance, procurement, overall production planning and market research, etc.	Efficiency gains from vertical integration of extraction activity (Esso Resources), manufacturing production and distribution (Esso Chemicals, Esso Petroleum) and strategic management (Imperial Oil).
	(ii) Distribution net based on minicomputers in Vancouver, Edmonton, Sarnia and Montreal (Links gas stations and oil terminals; 'vertical' links to Toronto and Calgary head offices).	Continuous reporting of final market transactions to refineries and head office management.	Lower costs and improvements in inventory control and delivery schedules. Improved market forecasting and cash flow management.
	(iii) Private satellite net (Links remote production sites in northern Alberta to Toronto and Calgary head offices).	Progress reports on exploration activity; production scheduling for 'downstream' (refining and manufacturing) operations.	Improved knowledge of supply conditions and lower inventory costs permitted by synchronised refining and manufacturing production schedules.

COMPANY DESCRIPTION	TYPE OF COMPUTER NETWORK	APPLICATIONS	REPORTED IMPACTS
	(iv) Radio links between Calgary head office and exploration sites in Arctic Canada.	Seismic data processing.	Greater efficiency in exploration activity.
	(v) Transnational data circuits between Imperial Oil's Toronto head offices and Exxon Corporation's regional offices in New Jersey, US (Parent-subsidiary link).	Financial and market reporting to US parent company.	Improved production scheduling for key US oil and gas market. Enhancement of Exxon's global integrated strategy/greater control over Canadian subsidiaries.

Source: Hepworth (1987a)

See also Appendix B

Some general comments

The case studies provide little evidence on the *local impacts* of computer network innovations, given the research's technical focus on 'backbone topologies' and its broad inter-regional perspective on possible changes in the geography of the enterprise. That is to say, my descriptive analysis of technological innovation (in most cases) treated corporate operations at each 'network node' as a 'black box'; similarly, the research neglected to consider whether or how local economies were affected by the technological changes underway in the firms studied. Clearly, then, these 'missing links' would need to be identified through research on innovation processes at the workplace level (flexible manufacturing systems in factory production, computerisation in offices, automated inventory in retail outlets and so on), and at the level of inter-firm linkages in specific local economies. In this respect, the case study findings may be seen as complementary to other geographical research on 'information technology and flexible production', which has tended to focus attention on the local rather than the inter-regional dimension of innovation processes (see Chapter Six).

The case studies, as a whole, point to the multifaceted role of computer networks as process innovations. In particular, these innovations alter the firms' technical production function but also the *control parameters* of production processes. Thus, while networks may function to support spatial decentralisation in productive, marketing, technical or management operations, their simultaneous use as 'control innovations' suggests that the locus of decision-making power within organisations could remain unchanged. As such, case studies of the impact of information technology on the firm would need to be in-depth and longitudinal investigations of changing organisational structure, behaviour and culture. Further, where computer networks are used to support inter-firm transactions, our analysis of the control attributes of information would need to take-in the entire 'filière' of productive and institutional (ownership-control) relations in which the innovation process is embedded.

The last aspect of the case study findings I wish to highlight relates to the internationalisation of business activity attendant to computer network innovations. In this respect, we can clearly see the development of a global information economy, based on commercial (traded) and corporate (traded 'in-house') transborder data flows (TDF). Importantly, these TDF evade national tariff and most non-tariff barriers, and they represent an international transfer of not just information services by also *capital*. That is to say, the 'communicability of information capital' (see Chapter Four) has an international dimension, which has gone unnoticed in the present GATT (General Agreement on Tariffs and Trade) talks on the 'modernisation' of commercial regulations in the world economy. Clearly, as geographers increasingly turn to consider new stages in the international division of labour, the role of TDF in foreign direct investment and trade merits considerable attention.

The case study findings summarised in Table 5.1 are, then, intended to provide some *indication* of the scope of computer network applications and their potential effects in large, multilocational firms. In the detailed studies which follow, we will encounter further evidence on these technological innovations which reveal the spatial implications more clearly.

The Globe and Mail *(newspaper publishing)*

The *Globe and Mail* newspaper began operating a satellite publishing network in October 1980 following two years of research and development (R and D) collaboration with the *Los Angeles Times*. The economic motive for computer networking, or change in production technique, was intensifying competition (from the *Toronto Star* and the

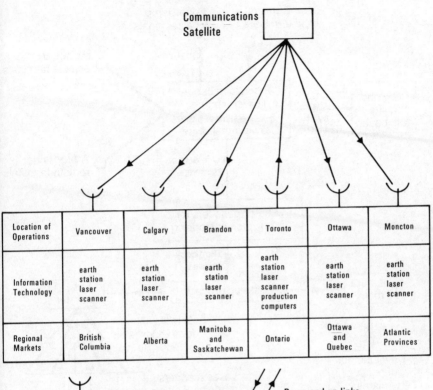

Location of Operations	Vancouver	Calgary	Brandon	Toronto	Ottawa	Moncton
Information Technology	earth station laser scanner	earth station laser scanner	earth station laser scanner	earth station laser scanner production computers	earth station laser scanner	earth station laser scanner
Regional Markets	British Columbia	Alberta	Manitoba and Saskatchewan	Ontario	Ottawa and Quebec	Atlantic Provinces

Y̶ Earth station   // Down and up links

Figure 5.1 *The* Globe and Mail *network*

Source: Hepworth (1987a)

Toronto Sun) in the *Globe*'s 'home' regional market of southern Ontario. Before the network (Figure 5.1) was established, the basic constraints on market expansion outside the 'home' region were high transporation costs (relative to printing costs and price) and the imperative of distributing the newspaper in time to meet morning commuter traffic in distant Canadian cities.

The satellite network is integrated into a highly computerised process of newspaper production at the *Globe*'s editorial offices and central printing plant in Toronto (Figure 5.2). In the final pre-printing stage of production, laser scanners are used to create an electronic facsimile of each edition, which is then broadcast to five 'remote' printing plants. Satellite signals are received simultaneously by all

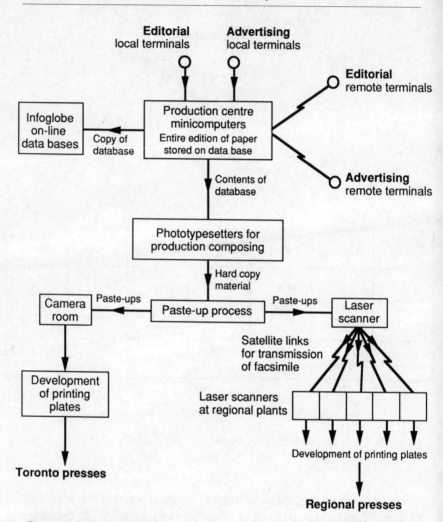

Figure 5.2 *The organisation of computer network-based newspaper production*

Source: Hepworth (1987a)

plants and converted to photographic negatives, using laser scanners, from which offset plates for the local presses are developed. This entire process is remotely controlled from Toronto, through the use of leased data circuits for scheduling the transmission of intermediate production inputs by satellite and for network maintenance (remote diagnostics).

The *Globe* is now published simultaneously across several time-zones and distributed from six printing sites to different regional markets. As a result, daily circulation of the newspaper outside Ontario has increased from 20 000 to 130 000 copies. Additionally, the average cost of 'exporting' the region to other regions has fallen by 25 per cent, owing to scale economies in satellite transmission. Wider circulation and a larger sized newspaper (compared with the light-weight 'export' copy dictated by transport costs) have acted to attract higher priced national advertising. In the longer term, market entry barriers have been created at the regional level, and further expansion of productive capacity at new remote locations will involve investments in only the ground segment of the network. The fundamental effect of these changed revenue and cost conditions is, therefore, that the *Globe* is Canada's only 'national' newspaper — that is to say, it now dominates the domestic newspaper publishing industry.

The spatial division of labour arising from this classic illustration of the telecommunications – transportation trade-off is characterised by what Massey (1984) calls the geographic separation of 'the functions of conception on the one hand and execution on the other' (p. 73). In the *Globe*'s case, 'conception' tasks, such as writing and editing articles, designing page formats and informatics management, are all carried out centrally, in Toronto. At the remote printing plants, by contrast, jobs have been created in traditional production occupations (e.g. pressmen), transportation (truck drivers for newspaper delivery) and in lower order informatics occupations (computer operators). It is, therefore, clear that a geographically dichotomised pattern of job-creation, whose qualitative variations reflect parallel variations in the new information capital deployed (traditional printing presses/computerised, pre-printing production techniques), had attended the *Globe*'s achievement of greater locational flexibility in productive operations and market expansion.

These case study findings are identical to the results of Bakis's (1982) research on the French newspaper industry. Other supporting evidence is reported by Irwin (1988: 35):

Satellite relay, facsimile equipment, scattered printing offices, for example, reduce the geographic constraints that limit a newspaper's coverage. *USA Today* is a classic example of a newspaper that combined information technology and moved beyond its location in Washington, DC. The *New York Times* and *Wall Street Journal* have become national newspapers as a result of remote facsimile, satellite relay and remote printing facilities. The global implications of remote printing are equally compelling. The *Wall Street Journal* and *USA Today* export newspapers to Europe and the Far East by satellite. At the same time the London *Financial Times*, and the *China Daily News* are imported to the US by satellite relay. The geographic reach of a newspaper is obviously being redefined by information technology.

We will return to the broader implications of this altered geographical reach permitted by computer network innovations in newspaper publishing.

Bell-Northern Research (electronics R and D)

When this case study was undertaken (in late 1984) Bell-Northern Research (BNR) accounted for about 60 per cent of R and D carried out in all Canadian manufacturing. The dominant users of BNR's software products are its Montreal-based parent companies, Bell Canada (telecommunications services) and Northern Telecom (electronics manufacturing). About three-quarters of BNR's R and D activity is embodied, as software systems, in Northern Telecom's main product lines — telephone switches (for public and private networks), circuit boards and office automation equipment. The company's head offices and main R and D laboratories are located in Montreal, but several other laboratories are dispersed across Canada, the US and, more recently, Europe.

The BNR computer network (Figure 5.3) was set up in 1977–78 to support an R and D process which has a complex functional and spatial organisation. A 'layered' approach characteristic to computer-based software development enables each laboratory to specialise in only part of the work carried out on a specific (Northern Telecom) product line. For example, base software and hardware for telephone switches are designed and engineered in Ottawa, but user-oriented application software for the same product is developed at 'remote labs' located in Texas and North Carolina (Figure 5.4).

Although this system of laboratory specialisation permits economies of scale in software production, its control requirements are

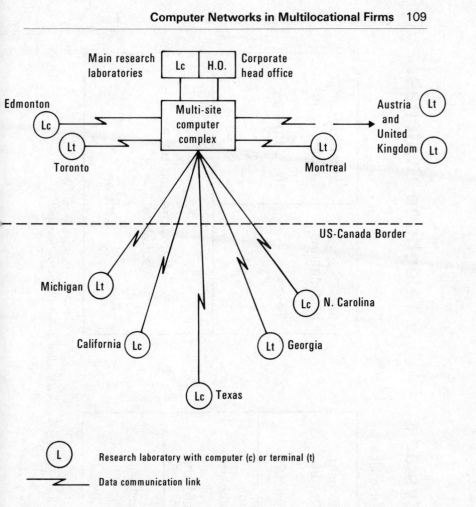

Figure 5.3 *The Bell-Northern Research network*

Source: Hepworth (1987a)

enormous, given that R and D teams must be kept perfectly informed of modular changes in product specifications that arise elsewhere. These control requirements increased greatly during the late 1970s, when new BNR laboratories were established in the US to support Northern Telecom's strategic shift into foreign markets, as domestic market growth slowed (Table 5.2). This spatial decentralisation of R and D activity is dictated by the need for face-to-face contacts between BNR's laboratory teams and corporate purchasers of Northern

Office Automation Systems

Small Users

end user software	Michigan
base software	California
hardware	Ottawa

Large Users

end user software	Toronto
base software	California
hardware	Ottawa

Telephone Switches

toll applications	local exchange applications
Texas	N. Carolina
hardware and base software	
Ottawa	

Figure 5.4 *Inter-laboratory specialisation in Bell-Northern Research*

Source: Hepworth (1987a)

Table 5.2 *Recent trends in Northern Telecom's sales and employment by geographic region*

	SALES ($ million)			TOTAL EMPLOYMENT		
	Canada	US	Total	Canada	US	Total
YEAR						
1982	1 248	NA	3 036	18 464	13 377	34 449
1981	1 335	1 047	2 571	20 776	12 737	35 444
1980	1 084	807	2 055	18 634	11 479	31 915
1979	1 001	740	1 901	18 511	14 147	33 301
1978	1 008	447	1 505	17 487	12 607	31 756
1977	1 014	193	1 222	18 303	4 048	23 577

Source: Evans Research Corporation, EDP In-Depth Reports (Toronto).

Telecom's manufactured products — for identifying user requirements, ensuring correct installation and post-sales support.

The BNR computer network has played a critical enabling role in this spatial decentralisation of R and D, pursuant to Northern Telecom's international market expansion. The complex logistical requirements of R and D activity are handled centrally (in Ottawa), with planning and co-ordination functions being supported by data base management systems and on-line progress reporting. In this technical context, the computer network also functions as a 'transportation' system for exchanging product-related information between remote laboratories and consolidating all of this information in central data bases (Ottawa). Before 1977–78, BNR relied on road and air transport for the physical delivery of computer tapes on which this information was stored. However, these modes of communication became increasingly inefficient at higher transaction frequencies as the scale and geographical dispersion of R and D activity increased. More and more special couriers were required physically to transfer computer tapes and supervise Canada–US customs inspection; more and more paperwork had to be prepared for customs authorities; and, the security risks of transferring R and D information grew accordingly. The computer network, in Rada's (1984) terminology, significantly enhanced the 'transportability' of finished and semi-finished software products, particularly because their embodiment in Canada–US transborder data flows (TDF) had the effect of eliminating the transaction costs associated with customs clearance.

The computer network is, of course, also used in the direct production of software through 'teleprocessing': most of the remote R and D laboratories still interact with host computer systems located in Ottawa to develop software products. Further, all of the computer systems used by the R and D laboratories are maintained through remote diagnostics carried out over data communications lines from Ottawa. This spatial on-line organisation of the production process permit economies of scale and scope to be achieved in information capital (computing/telecommunications resources) and in specialised information labour (higher order informatics personnel). At the same time, by centralising the informatics function, economies of scale are achieved in procurement, and head office management are able to monitor closely the R and D/software production process.

As a result of this network-mediated spatial division of labour, pursuant to Northern Telecom's foreign market expansion, the bulk of new R and D jobs (500 plus) created by BNR during the early 1980s was located outside Canada, and mostly in the US. The informatics labour force, used to support technically this more spatially distributed (software) production capability, grew only fractionally over this period (in Ottawa); at the remote laboratories, additional informatics jobs were trivial in number and involved 'machine-minding' tasks. Thus, this spatial division of labour is entirely consistent with the *Globe* study, namely decentralisation of productive activity (software development/newspaper printing) to enable regional market expansion, and, the geographical dichotomisation of job-creation in higher and lower order informatics occupations.

We already know, of course, that R and D activity in the electronics industry is relatively 'footloose' (Loinger and Peyrache, 1988), and that market proximity is a major factor contributing to spatial decentralisation at the international level (Milne 1989). Importantly, there is strong evidence of this locational flexibility being enhanced by the use of computer network technology in R and D (Howells, 1988). The particular ironic twist, highlighted by this case study, is that employment in Canada's leading source of manufacturing R and D 'migrated' to the US, at a time when some of the country's economic geographers (see, for example, Britton and Gilmour, 1978) were attributing Canada's diminished technological capabilities to R and D imports within foreign multinational companies.

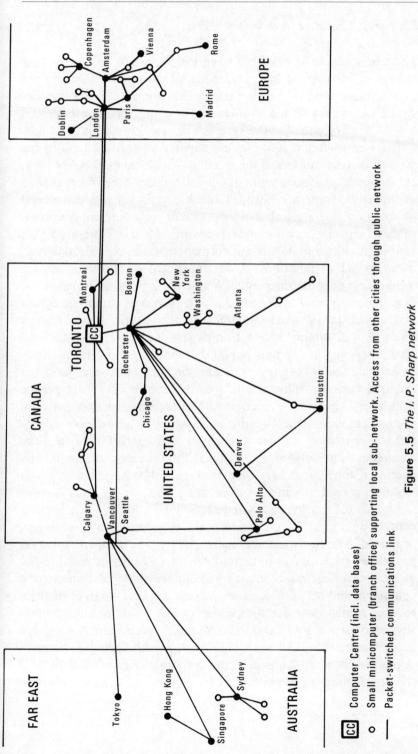

Figure 5.5 *The I. P. Sharp network*

Source: Hepworth (1987a)

CC Computer Centre (incl. data bases)

o   Small minicomputer (branch office) supporting local sub-network. Access from other cities through public network

—   Packet-switched communications link

I.P. Sharp Associates (computer services)

This case study was carried out when I.P. Sharp was an independent company, however, it has since been taken over by Reuters, the transnational services conglomerate. The company is a computer service bureau whose head offices and computer centre are located together in Toronto's central business district. Formed in 1964 as a software consultancy, it has used computer network technology for product diversification and the creation of a global market. By 1984, annual revenues amounted to about $50 million, over 70 per cent of which derived from sales outside Canada, and sixty branch offices and 600 personnel were distributed across twenty-two different countries.

Sharp's private packet-switched network, IPSANET (Figure 5.5), is used to produce and distribute informatics services for sale mainly to transnational corporations. These integrated technologies include computer processing time, systems and application software, public data bases, private data storage and network communications. Systems software is used through Sharp's time-sharing service and/or leased as a distributed product on customer computer installations. The company's ninety plus public data bases contain mostly time-series statistics on energy, aviation, finance, economics and the actuarial business. Information from this source, and from private data bases, is generally processed with retrieval and analysis software to carry out management-related applications — for example, econometric forecasting, project evaluation and market analysis. The communications network adds value to this processed information by making it timely and accessible to dispersed management users in multilocational firms and government agencies.

As a global distribution system, IPSANET's nodes are essentially delivery points for Sharp's products and services — that is, software, processed data, electronic mail messages, etc. The company's market geography can be inferred, therefore, from estimates of nodal traffic and connect time which measure customer usage of time-sharing computers and network communications services. Analysis of these revenue-related data (for an average business day in August 1983) suggested that Canada accounts for about 25 per cent of Sharp's global market, or half of the US share (Table 5.3). Within Canada, Toronto's dominance of national finance and corporate head office activity is reflected in the city's 64 per cent share of total connect time; elsewhere,

Table 5.3 *Distribution of data traffic and computer connect time for I.P. Sharp network*

MARKET AREA	NETWORK TRAFFIC (kilocharacters)		CONNECT TIME (hours)	
	Volume	Share Per cent	Time	Share Per cent
United Kingdom	22 579	10.0	719	12.0
Rest of Europe	16 681	7.5	601	10.0
Far East & Australia	17 603	7.5	569	9.0
United States	112 397	50.0	2 805	45.0
Canada	53 414	25.0	1 468	24.0
TOTAL	222 674	100.0	6 162	100.0
CANADA ONLY (100%)				
Toronto	29 768	55.6	9 44.5	64.4
Calgary	8 711	14.5	1 06.0	7.2
Ottawa-Hall	5 979	11.2	1 52.0	10.3
Victoria	2 450	4.6	57.0	3.9
Edmonton	2 012	3.8	65.5	4.5
Montreal	1 911	3.5	55.0	3.7
Vancouver	1 696	3.2	31.0	2.1
Winnipeg	754	1.4	30.5	2.1
Halifax	620	1.2	7.5	0.5
Saskatoon	249	0.5	11.5	0.8
London	164	0.3	7.5	0.5

Note: The source data were obtained in the form of a computer printout, showing traffic and connect time by individual network nodes.

Source: Hepworth (1987a).

Sharp's principal clients are the federal and provincial governments, multinational oil companies in Alberta and Montreal-based airlines.

The spatial organisation of network-mediated production can also be inferred from IPSANET's topology, in combination with the geographical distribution of informatics personnel (Table 5.4). Although the computer centre is located at head office, 56 per cent of the company's personnel work outside Canada, interacting with the Toronto systems over the network. Distributed processing, or decentralising of this productive capital, is not made feasible by the size of

Table 5.4 *Location and work functions of I.P. Sharp informatics personnel (June 1984)*

NAME OF GROUP	LOCATION OF PERSONNEL	GROUP FUNCTIONS
Application Software	Rochester, US = 4 Chicago, US = 2 Sydney, Australia = 2 Ottawa = 4 TORONTO = 27 TOTAL = 40	Development of packaged application software
Research and Development	Rochester, US = 6 Palo Alto, US = 5 TORONTO = 20 TOTAL = 31	Development, installation, and maintenance of SHARP APL systems software
Public Data Base Services	London, UK = 5 Sydney, Australia = 5 Far East = 3 TORONTO = 25 TOTAL = 38	Creation and maintenance of data bases, incl. access software; data acquisition, assembly and transmission
Special Systems Division	Variable locations. In May 1985, personnel were in Boston, Palo Alto, and Amsterdam TORONTO = 60 TOTAL = 70	Specification and implementation of on-site computer systems
Computer Network Operations	All in TORONTO, incl. computer operators at data centre and network operations group TOTAL = 30	Operate and maintain network for in-house and commercial services
Branch Support	Rochester, US = 25 London, UK = 25 TORONTO = 30 Remainder are evenly distributed between branch offices (about 4 to 5 per branch) TOTAL = 350	Adapt standard application products for local users; technical support and user training; local market reporting; advertising of new products

Source: Hepworth (1987a).

Sharp's current market, given the high costs of duplicating mainframe-based, time-sharing operations (with back-up systems) and updating data bases (a labour-intensive process) at separate locations. Thus, IPSANET permits economies of scale and scope in the production of informatics services by enabling Sharp personnel at all locations to share central computing resources 'over the wire'. For example, while public data bases are constructed, updated and maintained in Toronto, part of the production process is decentralised to other cities. Local staff obtain data from 'information providers' (e.g. national stock exchanges), and prepare and forward these data over IPSANET for final distribution from Toronto. The need for decentralisation arises, in general, from Sharp's marketing strategy of supplying a complete range of services to transnational corporations whose information and communication requirements are global in scope.

For similar reasons, software development is also a centralised and partially decentralised process. Scale economies are achieved by concentrating three-quarters of the Application Software group in Toronto, but marketing of main product lines requires face-to-face contacts with a dispersed customer base. Consequently, about 70 per cent of branch staff are based outside Canada, using IPSANET to adapt products for clients, provide technical support, disseminate new product information and report local market developments. Software development is also carried out in remote cities by the company's Special Systems Division and Research and Development group. These personnel, who also interact with the Toronto computers, work closely with individual clients (e.g. a Chicago bank) in developing new software products and implementing on-site computer systems. For example, in May 1984, seven customers leasing Sharp's software system for manufacturing 'microchips' were located in California's Silicon Valley (Palo Alto).

The recent decline of its general time-sharing markets, which is attributed to greater international competition and the growing internal use of small computers by clients, led Sharp to scale down network operations. Rationalisation measures announced in April 1985 reduced the number of branch offices from sixty to forty-nine, and the workforce from 600 to 530. The economic motives for closing Canadian branches in London and Winnipeg, for example, are suggested by the usage data presented here (see Table 5.3). The company has also repeatedly publicised the possibility of re-locating its head offices and computer centre to the US, where computer and telecommunications costs are up to 30 per cent lower and about half of its existing market (1984) is concentrated. Technical constraints on any spatial reorganisation of production are, indeed, minimal owing to the high level of distributed control built into IPSANET and the feasibility of upscaling regional operations in Rochester, New York.

Four years later (August 1988), the company is still rooted in Toronto, although this may (or may not) change with the Reuters takeover and the recent US–Canada 'free-trade' pact in computer services. In mid-1984, successive public statements on relocation possibilities were essentially political threats, aimed at the Canadian federal government, in an attempt to have tariffs and taxes on computer equipment lowered. However, computer equipment and telecommunications services, respectively, account for only 15 per cent and 8 per cent of Sharp's total operating costs, while salary payments make up over 50 per cent. With electronic mail systems keeping clerical and secretarial employment to a minimum (10 per cent of the workforce), the bulk of this wage bill is tied up in human capital investments. Thus, the most credible explanation (endorsed by interviewees) for Sharp's apparent locational inertia is what Williamson (1975) calls 'human asset specificity', as it applies to the company's Toronto-resident staff of informatics specialists.

This case study, then, illustrates what Dordick, Nanus and Bradley (1981) call the 'network marketplace', wherein information is bought and sold as 'on-line' commodities. We know very little about the geography of this 'marketplace', except for its dominance by the US in the context of services trade (Sauvant, 1986).

Competing in Time

Competing in Time is the title of Peter Keen's (1986) book on the business management uses of telecommunications. Like other books of this type (see, for example, Porter, 1985; Wiseman, 1988), it suggests how private computer networks may be used to enhance effectiveness and efficiency in large corporations. Much of this business literature can be found in the popular trade press. The following discussion is organised around an assortment of this rich, anecdotal, material.

My own case study research did not, for example, extend to

videoconferencing applications of computer networks. Consider, then, the following example of Chrysler Corporation's use of this class of network application based on satellite communications:

> Between Chrysler and its [6000 domestic/US] dealers will bounce car and parts orders, information for service technicians, and sales information such as loan pay-offs and credit applications and approvals. The network will also beam video training classes, information programming, and Chrysler's news shows to dealers from headquarter studios in Highland Park, Michigan.[3]

Several of the Canadian case studies have already illustrated these types of network applications (order entry, inventory control and credit transactions) and their rationale. The Chrysler example does, in addition, highlight a specific area of microeconomic analysis which is concerned with principal–agent problems (Strong and Waterson, 1987). These problems arise when one party to a contract (the principal: Chrysler) engages another party (the agent: Chrysler's dealers) to carry out actions on its behalf (sell cars) in situations of asymmetric information (the dealer's greater knowledge of local market conditions). In this context, where individual dealers may pursue their own interests (say, by playing off one car manufacturer against another in sales promotion), there is an incentive for Chrysler (say) to invest in 'control technologies' (satellite communications) and to offer incentives (improved dealer-support services) which permit monitoring of dealers' actions and their supplier loyalty to be enhanced.

More generally, the rationale for Chrysler's use of satellite media is to increase the information efficiency of the company's 'control space', as dictated by the geography of its mass consumer market and mass distribution system. As such, the Chrysler network is the latest information 'control technology' used for this purpose, following in the long historical line of innovations identified by Beniger (1986) (See Figure 1.1). The point-of-sale network used by Sears Canada, of course, also falls into this innovation category (see Table 5.1).

A further growth area of teleconferencing applications is 'trouble-shooting' and remote 'expert systems' in research and product development. The following example illustrates the US Bechtel Group's use of the technology for supporting international engineering projects:

If a construction problem arises, on-site engineers can take Polaroid snapshots or shoot videotape, which would be transmitted to wherever the experts are via very small aperture satellite dishes. Bechtel plans to connect job-site CAD [Computer-Aided Design] units with those in offices, so engineers at each end of the problem can look at the same information concurrently. Today, drawings are sent by facsimile. As soon as economics permit, the CAD systems will be connected [by satellite transmission].[4]

Thus, expert systems applications of computer networks may 'collapse space and time' in R and D. Whereas the BNR study highlighted the production flexibility aspects of network-mediated R and D, permitted by economies of scope in the shared (inter-lab) use of informatics resources, the Bechtel example, on the other hand, draws attention to the economies of scope in construction engineering project work permitted by the diffusion of more 'transportable' technological know-how. While both of these sources of economies of scope may encourage spatial decentralisation tendencies in productive operations, where this is dictated (directly or indirectly) by the geography of market demand, it is extremely plausible that R and D activity itself will be fragmented into smaller scale, more specialised units with variegated locational propensities. This possibility, and its potential effects on localised 'technological clusters', is suggested by Loinger and Peyrache (1988: 126–7):

> In this context, the transmission of data becomes important; the technology of video-screened work is naturally used to the full, with daily video consultations between units on the methods bureau level. At the same time, the sophisticated organisation of long-distance conception systems and especially computer-assisted production increases the number of communications 'networks'. One could ask whether it is possible effectively to diffuse technological innovations locally and regionally from this kind of structure: the separation of R & D poles is not favourable to the development of local and regional contexts.

There are numerous reported cases of private computer networks being used to support the internationalisation of markets. The following example illustrates Citibank's use of automated teller machine (ATM) applications for international expansion:

> To help penetrate foreign markets, Citi[bank] offers as bait access to its US ATM network. The Tokyo newspaper *Yomiuri Shimbun* reported that

Japan's largest bank, Dai-Ichi Kangyo Limited, and Citi agreed to let each bank's customers use ATMs in the other bank's country. In addition, Dai-Ichi agreed to support Citi's MasterCard in Japan. From Citi's point of view, the move helps overcome barriers the Japanese have erected to prevent foreigners from establishing full-service consumer banks on their islands.[5]

This instance of joint marketing agreements in financial services, as it applies to international ATM networks, was also evident in the CIBC case study (see Table 5.1). In the airline industry, there are similar market alliances developing around the competitive use of reservation systems, as the Air Canada study indicated. In the Asia–Pacific region, for example, Cathay Pacific, Singapore Airlines and Thai International have established a joint venture company to operate a computerised reservation system called 'Abacus', in order to maintain their market share against European and American competition. In this context, reservation systems have clearly emerged as a 'competitive weapon':

Development of a regional system in the Asia–Pacific area is vital because the region is the world's big growth travel market, and whoever controls the reservation system will make the biggest profits. These so-called mega-reservation systems cost millions of dollars to develop and operate. But, they are not simply an information system. They are a completely automated travel service that already has revoluntionised world travel.[6]

The last two examples underline the central role of transborder data flows (TDF) in the internationalisation of service industries (Sauvant, 1986). Only a few economic geographers (for example, Howells, 1988; Bakis, 1987; Langdale, 1987) appear to be taking significant account of TDF, despite its profound importance for the global restructuring of industries and markets. Intra-corporate TDF are no less important in the manufacturing sector, where multinational firms are using computer-aided design and manufacture technology over high-speed digital links to lower production costs and improve product design (Cohen and Zysman, 1987). The following example relates to other computer network applications in manufacturing:

For example, Hewlett-Packard, the electronics firm, operates 53 manufacturing plants worldwide. These facilities link to HP's global telecommunications network. To process each plant's orders, HP developed a centralised electronic procurement system. From this SIS [Strategic Information System], it reaps economy-of-scale benefits, obtaining more favourable pricing and delivery terms from its vendors. HP uses information technology here to support a global cost thrust targeted at the supplier arena.[7]

As several of the Canadian studies indicated, private networks also play an important role in physical distribution logistics — for example, the scheduled movement of retail merchandise and gas station supplies in the Sears and Imperial Oil cases respectively (see Table 5.1). In the road haulage industry itself, network systems are increasingly used to develop a 'competitive edge' in shipment markets, as this press extract reveals:

Back in 1971, Pacific Intermountain Express, a large trucking firm based in California, developed an on-line application for tracking the status of a shipment and freight billing. PIE developed an application that lists all inbound shipments due on its trucks. PIE raised the information systems ante for other trucking firms. But it wasn't enough. Consolidated Freightways and Leaseway Transportation have matched PIE's investment and now feature information-based services to distinguish themselves from other 18-wheelers on the road. In 1984, Consolidated installed an on-line system called Direct, which provides critical data to more than 1,500 shippers on the progress of shipments. Big customers like American Hospital Supply get specialised reports. In 1985, Leaseway advertisements proclaimed a technological revolution in transportation, including electronic routing and scheduling. Clearly, PIE didn't win the strategic information war. What they offered was too easily duplicated: the barriers to entry weren't barbed-wire topped.[8]

The last example highlights the growing significance of information services to market competitiveness within different sectors, including manufacturing industry. Over the last decade, these 'quasi-markets' for information services have been increasingly 'externalised', as part of what Lemoine (1981) calls the 'informatisation' of the economy (see also, Miles, 1988a; Ciborra, 1983). Elsewhere, Hepworth (1989), I have interpreted these organisational transformations as signifying the 'redevelopment' of information space (Tornqvist, 1974; Thorngren, 1970) through commoditisation, electronification and internationalisation. The case study of I.P. Sharp, the computer services bureau, exemplified the commercial activities of firms involved in this 'redevelopment' process.

Finally, let us draw on a press-cutting which serves to illustrate further the complex spatial divisions of labour made possible by computer network innovations. Perhaps this is best shown by the Caribbean data entry industry, which a growing number of US companies has established to take advantage of a cheap, non-unionised and largely English-speaking workforce. Consider, for example, Jamaica's planned development as an offshore production centre in the global information economy:

> Jamaica is counting on its long-awaited teleport to boost business there. Once its modern phone links are in place, local data-entry companies will be able to transmit their finished work electronically instead of using unreliable and costly air service to send computer tapes back to the US. At first the telecommunications service will be available only to companies in the Montego Bay free-trade zone. But, it will be extended – first to Kingston and other free-trade zones, and eventually to data-entry companies throughout the island. Jamaican officials are also hoping that the new telecommunications link will lure other information-service businesses to Jamaica. For example, toll-free telemarketing services such as airline reservations or customer assistance could be handled there. Jamaican and American Telephone & Telegraph Company officials are predicting that such work could create 10,000 jobs for the island within five years.[9]

These emerging developments may, according to Pelton (1986), signify a new basis for the Third World's incorporation into the international division of labour — as 'electronic colonies' for low-value information processing (Cruise O'Brien and Helleiner, 1983). But, more generally, the example illustrates the potential uses of computer network technology for systematically exploiting low-wage, non-unionised labour markets (Hepworth and Robins, 1987).

In this case, locational flexibility in the production process provides firms with greater scope for instituting flexibility in the labour process, in terms of concessionary adjustments to the employment relation (Bluestone, 1987). It is extremely plausible, therefore, that the balance of power in collective bargaining will shift significantly in favour of employers, particularly as large firms progressively make instrumental use of production and locational flexibility to transcend local concentrations of trade union influence (Aglietta, 1979). In the UK, for example, distributed production systems identical to those computer network techniques used by the *Globe and Mail* newspaper (see Figure 5.2) have greatly diminished the bargaining status of traditional printing chapels in London's Fleet Street district.

Organisational and spatial tendencies in the 'network firm'

I have made fairly liberal use of press-cuttings, and the Canadian case studies by themselves are obviously limited in number and by their depth of issue analysis. However, in a research area where both theoretical and empirical work is in its infancy stages, and where access to corporate data on the 'nervous system' of the firm calls for some degree of pragmatism, there is good cause for proceeding to generalise on a tentative basis. What follows, therefore, are a few cautious generalisations on spatial tendencies in what Antonelli (1988b) calls the 'network firm'.

The 'network firm', according to Antonelli, is the result of the differentiated effect of 'telematics' (computer networks) on the production function and the governance function, with the latter being the set of organisational techniques available to govern transactions between production units (see, for example, Williamson, 1985, 1986a). On the basis of several case studies of telematics usage in large Italian firms, Antonelli suggests that:

> The effects of microelectronics on production functions tends to be basically centrifugal, reducing the role of technical economies of scale by lowering the minimum efficient size of production unit. The effects of telematics on governance functions, instead, appear to be rather centripetal, increasing the weight of economies of scope and the efficient size of governance structures. The overall result of such a conflicting mix of centrifugal and centripetal forces generated by the introduction of telematics is a growing trend towards the appearance of a new institution: the 'network firm' characterised by the large size of the governance structure and small size of separate production functions and business units managed by means of hierarchical organisation and quasi-integration. [p. 21]

Antonelli's generalisations on the 'network firm', which are consistent with Camagni and Rabellotti's (1988) findings on the Italian textile sector, flow logically from our earlier theoretical discussion on the potential effects of computer network innovations. They are, in addition, broadly consistent with the case study findings for my sample of large Canadian-based firms.

We saw from the detailed case studies of the *Globe and Mail*, Bell-Northern Research and I.P. Sharp, in particular, that network innovations permit firms not only to decentralise productive capacity but also to enlarge their 'span of control' in a spatial context. While these limited findings tend to suggest that network innovations may encourage the further development of 'hierarchies' rather than 'markets' (Williamson, 1975), we should bear in mind that the case studies provide only a 'snapshot' of the embryonic stages of technological and organisational innovation in the firms sampled. Also, in the case of relatively new industries, or where fundamental process innovations, (such as computer network) are still immature, the transactional model of the firm holds that 'information impacted-ness' problems will favour 'hierarchies' (internalisation) rather than 'markets' (externalisation) — see Silver (1984). However, in the longer term, it is extremely plausible that this market hierarchy balance of transactions will change in the Canadian firms sampled, as Stigler (1951) and Williamson (1986c) suggest, leading to a more complex, alternative, set of organisational and spatial dynamics (Scott and Angel, 1988; Nicholas, 1986).

We have also encountered some evidence of what Antonelli calls 'electronic quasi-integration', that is: 'contractual agreements between independent firms based on a shared computer communications infrastructure' (p. 24). This form of quasi-integration, permitting external economies of scale in information-related activities or 'network externalities' (Wilson, 1975), was illustrated, for example, by CIBC's ATM network and its use of the inter-bank SWIFT network, and also by Air Canada's reservation system and its use of the inter-airline SITA network (see Table 5.1). In both of these cases, network externalities are exploited to lower the operating costs of international market expansion.

Additionally, Air Canada's resale of it data base services (reservation system) to other related sectors (travel agents, hotels, etc.) illustrates forward-linkage developments which Antonelli (1988b) dubs 'electronic franchising'. These 'franchising' arrangements, which parallel the Chrysler example of satellite communications-based manufacturer-dealer relations, enable Air Canada to achieve greater control over distribution and marketing at decreasing managerial co-ordination costs. What these particular examples suggest is that the geographical field of 'franchise' relations is potentially enlarged, thereby highlighting the possibility of higher levels of internationalisation in retail chain operations (e.g. Sears) — see, for evidence, Wiseman (1988).

The informational economies of scale and scope, permitted by intra-corporate TDF, will clearly alter the balance of 'location-specific endowments' of countries and the 'ownership-specific endowments' of enterprises — in favour of the latter — which Dunning (1977) sees as the principal determinants of competitiveness in transnational corporations. While this may encourage, as the Canadian studies suggest, a pronounced propensity of firms to internalise activities across national boundaries, it is extremely plausible that inter-organisational computer network innovations will have 'externalisation' effects.

In particular, what Dunning calls 'structural imperfections' and 'cognitive imperfections' — that is, the costs of using non-competitive markets under conditions of uncertainty — may be partly offset by 'electronic subcontracting' (Antonelli, 1988b) with local suppliers and the purchase of variegated on-line information from firms in the global 'network marketplace', such as I.P. Sharp. The former innovation, 'electronic subcontracting', permits opportunities for locking-in suppliers and appropriating economic rent through asymmetrical contractual relations, while lowering market transaction costs (Klein and Leffler, 1981); appropriate purchases in the 'network marketplace', on the other hand, act to increase the overall 'information efficiency' (Arrow, 1985) of both internal and external transactions. The ultimate effect of these organisational and technological innovations may be, then, to increase local procurement levels amongst foreign subsidiaries of transnational corporations, if the costs of using the price mechanism are reduced by more than the falling costs of organising transactions internally (Rugman. 1981).

Most of the Canadian studies highlighted the efficiency gains which accrue from the use of computer networks for functional integration, both within and across firms (see also, Camagni, 1988). A major result of these innovation processes, as suggested by the I.P. Sharp and IBM Canada cases, may be to reduce levels of 'bureaucracy' (or increase 'X' efficiency) throughout the large multilocational firm, but particularly in ancillary regional or branch offices. The elimination of redundancy in existing organisational structures may, at the same time, provide the necessary scope for higher levels of specialisation in management, technical and professional functions in a decentralised spatial context (Antonelli, 1988a). As such, we may be witnessing the emergence of what Drucker (1988) calls the 'information-based organisation' in the corporate sector, that is: a large corporation comprised of specialists and whose technical division of labour is more akin to that of a 'symphony orchestra', rather than to that of a large, pyramidal, bureaucracy.

Conclusion

In this chapter, I have been looking at the *early* stages of computer network innovations in large firms, and some of their attendant spatial ramifications. The case study evidence should, therefore, be treated as indicative of organisational and spatial tendencies whose future trajectory is not yet clearly visible. However, even on the basis of this limited compilation of case study material and press-cuttings, it is reasonable to surmise that private network innovations will have a profound impact on the geography of markets, production and employment in advanced industrialised economies.

What emerges from the available evidence, more specifically, is a complex and unfolding set of organisational and spatial dynamics characterised by locational flexibility in production and labour processes, a wide variety of inter-firm relations supported by so-called 'electronic integration', and changing market geographies at a national and international scale. These organisational and spatial tendencies, of necessity, will transform capital–labour relations as much as inter-firm relations within and across markets, and trade–investment relations in the national and global contexts. The discussion in this chapter is, then, no more than an initial attempt to assemble some ideas on the economic and technical processes underlying these expected transformations.

We have, therefore, begun to explore the information space of production which is evolving through computer network innovations at the level of the firm. This type of space is articulated by information capital flows and information service flows which have the potential to change the spatial and functional organisation of production in the individual firm and between contracting firms. With the penetration of information technology into factory operations (e.g. flexible manufacturing systems), and trends towards functional integration, computer networks are emerging as the key technologies for achieving economies of scale in the use of information capital and information labour at the level of the firm and networks of contracting firms.

The next chapter continues with this exploration of information space, and considers its gathering significance for geographical analysis of 'flexible production'. This involves looking more closely into the informational structure of firms and inter-firm relations in a regional context, as well as the role of information capital in transforming production processes in manufacturing industry.

Notes

1. OECD (1983). A Survey for the Business and Industry Advisory Committee (BIAC) and the OECD. Discussion paper, OECD Symposium on Transborder Data Flows, London, December.
2. Informatics occupations (as defined by Statistics Canada) include: data entry operator, data control clerk, machine operator and computer operator (i.e. lower order 'Operations' staff); systems analyst, application programmer and methods and procedures analyst ('Applications'); systems software programmer, hardware specialist, minicomputer specialist, telecommunications specialist and graphics specialist ('Plant'); user trainer ('Staff').
3. J. McPartlin and C. Hord, 'More firms turning to satellites', *Information Week*, 4 April 1988, p. 15.
4. P. Schindler, 'Bechtel's mission: shrinking time and space', *Information Week*, 2 May, 1988, pp. 31–2.
5. C. Wiseman, 'Strategic information systems', *Information Week*, 9 May 1988, p. 37.
6. T. Ballantyre, 'Risk to Qantas in delay over system choice', *Sydney Morning Herald*, 28 December 1987, p. 12.
7. C. Wiseman, 'Strategic information systems', *Information Week*, 9 May 1988, p. 35.
8. R. Layne, 'Guerrilla warfare: strategic IS is the real world', *Information Week*. 9 May 1988, pp. 40–1.
9. G. De George, 'The Caribbean: a back-office paradise', *Business Week*, 11 April 1988, pp. 84D–E.

Chapter Six

'Flexible Production' in Information Space

Introduction

We saw, in the last chapter, that computer network-mediated information linkages are increasing the locational flexibility of large multinational firms. This new layer of information space is founded upon other 'layers', created by a long succession of 'control innovations' (Beniger, 1986) and gradual extensions and refinements to the division of information labour (Porat, 1977). Indeed, without these past rounds of information capital investment, the modern corporation and the 'Fordist' mass production economy could not possibly have emerged. But, what of information space today, and the 'sloganisation' of 'flexibility'?

This chapter continues to explore the importance of information linkages for economic geography. Attention is given to two main areas of analytical interest which have lately risen to prominence in the field of industrial geography. First, and most obviously there is growing interest in the spatial implications of computer network innovations which are currently transforming *direct* production processes in manufacturing industry — new technologies such as 'flexible manufacturing systems' and 'just-in-time' systems. The relevant information linkages, here, are embedded in the 'production line', at the level of factory operations carried out within and across manufacturing firms. And second, of related interest are the information linkages which articulate exchange or 'transactional' relations within and between firms, indirectly defining alternative forms of industrial organisation. Interest in these dual information linkages, as most readers will know, lies behind the intensification of research work on so called 'flexible production systems' and their economic geography.

The purpose of this chapter, along with the ideas presented in earlier ones, is to offer my own perspective on these recent developments in manufacturing industry. It is, therefore, only another contribution to a burgeoning debate on the present changes in the character of production systems and their locational behaviour. Given the important role assigned to vertical (dis)integration and (derivatively) transaction costs in shaping the 'new' geography of manufacturing, I begin with a critical overview of some relevant theoretical literature on these economic concepts. Following this, I highlight the growing information intensity of manufacturing industry, a trend whose aggregate dimensions were partly revealed by Chapter Two's analysis of recent (and long-term) changes in the occupational structure of the labour force. Finally, my discussion of the spatial implications of strengthening information linkages in manufacturing concentrates on the relatively new work relating to 'flexible production systems'.

Coase and transaction costs

It is now fully half a century since the publication of Ronald Coase's seminal article on 'The nature of the firm' (*Economica*, 1937). In the interim, and following from the work of Robinson (1969), for example, on imperfect competition and the empirical studies by Berle and Means (1932) and Hall and Hitch (1939), for example, on corporate form and behaviour, economists have attempted to model the real-world complexity of the firm's motivational forces and its decision-making processes under conditions of uncertainty. Flowing from this line of research are managerial theories (for example, Baumol, 1962; Marris, 1964) and behavioural theories (for example, Cyert and March, 1963; Simon, 1955) of the firm. The economics of information, as it is concerned with information-related market failures, has also emerged from these critiques of orthodox neo-classical theory (see, for a review, Spence, 1974; Stiglitz, 1985; Lamberton, 1983).

The work of Coase (1937, 1960), in this regard, is more fundamental, owing to its 'heretical' questioning of the *raison d'être* of firms as mechanisms for co-ordinating resources in the economy: Why do firms exist? In developing an answer to this question, Coase, like Robertson (1923) before, elaborates the role of alternative mechanisms for co-ordinating the allocation of productive resources:

Outside the firm, price movements direct production, which is coordinated through a series of exchange transactions on the market. Within a firm, these market transactions and in place of the complicated market structure associated with exchange transactions is substituted the entrepreneur coordinator, who directs production. It is clear that there are alternative methods of coordinating production. [1937: 333]

On the basis of this differentiation of resource allocation systems, Coase suggests that 'the main reason why it is profitable to establish a firm would seem to be that there is a cost of using the price mechanism' (p. 336); thus, in offering an economic definition of the firm, he states that 'the distinguishing mark of the firm is the supersession of the price mechanism' (p. 334). Transaction costs are, then, generally understood to be the costs of using the price mechanism — that is, they are the resource expenditures associated with using the market to transfer a good or a service from one party to another.

As previously stated (see Chapter Two), Coase's definition of transaction costs emerges from the following statement:

> In order to carry out a market transaction it is necessary to discover who it is that one wishes to deal with, to inform people that one wishes to deal and on what terms, to conduct negotiations leading up to a bargain, to draw up the contract to undertake the inspection needed to make sure that the terms of the contract are being observed and so on. [1960: 15]

In dissecting this paragraph, Dahlman (1979) distinguishes between different phases of the exchange process, and concludes that transaction costs can be said to incorporate search and information costs, bargaining and decision costs, and policing and enforcement costs. He adds that:

> Yet, this functional taxonomy of different transaction costs is unnecessarily elaborate: fundamentally, the three classes reduce to a single one — for they all have in common that they represent resource losses due to lack of information. Therefore, it is really necessary to talk only about one type of transaction cost: resource losses due to imperfect information. [p. 148]

Dahlman's interpretation of the underlying source of transaction costs is, indeed, consistent with Coase's (1937: 338) view: 'it seems probable that a firm would emerge without the existence of uncertainty'. It follows, invoking Arrow's (1984) conception of 'information [as being] merely the negative measure of uncertainty' (p. 138), that the resources set against transaction costs by firms are basically informational. According to Goldberg (1985), for example:

The phrase 'transaction costs' captures the notion that transacting — engaging in economic activity — requires the use of real resources. It embodies two very different meanings. One focuses on identifiable activities involved in transacting. The concept would presumably include the costs associated with bargaining, negotiating, and monitoring performance — costs usually associated with the activities of purchasing agents, lawyers, accountants and similar functionaries. It is analogous to the Marxist concept of 'non productive labour'. [p. 399]

In Chapter Two (Table 2.12), data were presented from Wallis and North's (1986) quantitative study of secular trends in the monetised 'transactional' sector of the American economy, and we noted the apparent high degree of correlation between these expansionary trends with the long-term growth of information occupations in the economy. The rationale for making these *informal* sectoral comparisons is obviated by Dahlman's interpretation of transaction costs as information-related expenditures (see also Cheung, 1983). Similarly, Arrow's broad definition of transaction costs as the 'costs of running the economic system' 1969: 48) is consistent with this view of the wider forces underlying the growth of the information economy. He states elsewhere that:

Clearly, firms do engage in information gathering. They spend resources on engineering and market research. Moreover, there are large and significant exchanges of information through the market-newspapers, business advice, and in a somewhat modified sense, all of education — in short, the whole realm of the production and distribution of knowledge, which Fritz Machlup (1962) has so carefully measured. Thus, information is not merely a good that is desired and acquired but is to some extent a commodity like others whose markets we study. [1984:142]

While the concept of transaction costs is helpful to developing an explanation for sectoral change (by occupation and by industry) in the economy, it is more faithful to Coase (and Wallis and North) to share his primary emphasis on the role of these types of costs in shaping the institutional arrangements of the economic system. In particular, Coase's analysis was directed at explaining the different alternative institutional arrangements which arise from transaction cost-economising behaviour in firms.

In this regard, Coase (1937) attempts to explain the extent of a firm's activities — its level of vertical integration — with respect to the relative costs of organising transactions within the firm and across markets. He states:

A firm becomes larger as additional transactions are organised by the entrepreneur (rather than coordinated through the price mechanism) and becomes smaller as he abandons the organisation of such transactions. The question that arises is whether it is possible to study the forces which determine the size of the firm. Why does the entrepreneur not organise one less transaction or one more? [p. 339]

The well-known answer Coase offers is, of course, that the firm's size will be determined such that the costs of internalising an extra transaction will be equal to the costs of organising that transaction in a market or in another firm. Of particular interest is Coase's discussion of the factors that create a 'brake' on the firm's expansion. He argues that 'as a firm gets larger, there may be decreasing returns to the entrepreneur function, that is, the costs of organising additional transactions within the firm may rise' (1937: 340). As Blair and Kaserman (1983) have pointed out, Coase's explanation of the limits to firm size is based on the notion of managerial diseconomies of scale (i.e. the bounded rationality of managers), which he sees as being sensitive, *inter alia*, to the spatial distribution and diversity of transactions. Williamson (1973) offers a similar explanation, as noted in Chapter Five, suggesting that 'spans of control can be progressively extended only by sacrificing attention to detail. Neither transactional economies nor effective monitoring can be achieved if capacity limits are exceeded' (p. 323). We have noted already that both Coase and Williamson highlight the capabilities of 'new' information technologies and new organisational forms (say, the 'U' and 'M' form corporation) — as control innovations — for increasing the efficient size limits of the firm.

We should bear in mind that Coase's work is a blend of ideas from economics, organisational theory and law. The legal aspects of transaction cost economics is highlighted by Williamson's (1986c) remark that 'it reduces essentially to a study of contracting — which means that the contracting expertise of lawyers developed in other contexts can be drawn upon' (p. 197). Thus, in interpreting Coase's ideas on the firm and industrial organisation, Cheung (1983) comments:

> It is not quite correct to say that a 'firm' supersedes 'the market'. Rather one type of contract supersedes another type. Coase's main concern is a type of contract under which an input owner (say, a firm or a worker) surrenders a delimited set of rights to use his input in exchange for income. He is therefore directed by a visible hand, not by the invisible hand of a price mechanism. It takes remarkable insight to see that as this type of contract increases there will be fewer product markets. [p. 10]

The influence of Coase's analysis of transaction costs and vertical integration has grown through the 1970s and the 1980s. It should, however, be noted that his analysis is generally acknowledged to offer 'only' a good starting point for developing an understanding of the existence of different organisational forms and, in particular, how economic activity is divided up within and between firms. For example, Coase's limited treatment of the sources of transaction costs tends to undermine predictive analysis of what types of transactions are transferable across markets and how this might be expected to change. Coase's work represents, then, a preliminary, but path-breaking attempt to develop a framework rather than a detailed analysis of transaction costs, which Williamson (see below) and others have taken further.

Vertical integration and distintegration

The economics literature on vertical integration is made up of several theoretical contributions, although the 'market failure' framework identified with Coase and Williamson is presently dominant. In this section, I briefly review these different approaches to vertical integration, in order to highlight the broader explanatory field wherein transaction cost economics is situated. Attention focuses on the rationale for vertical (dis)integration based on 'efficiency' arguments

— relating to production technology and organisational form — and on models of strategic behaviour.

Most economics textbooks restate, albeit briefly, the technological interdependence argument for vertical integration. My motive for restatement is that it has obvious relevance for understanding the logistics and geography of so called 'just-in-time' manufacturing production (Toone and Jackson [eds] 1987; Estall, 1985). The familiar example of these technological externalities is the thermal economies (savings in fuel costs) obtained by siting various stages of metal manufacture in the same plant or in adjacent plants; locational concentration also provides considerable savings in transport costs. Williamson discusses examples of these spatially determinate 'interdependencies' under the rubric of 'site specificity'. He argues (1986c: 208–9) that these cost savings could be achieved by allocating production stages to separate firms, however, transactional difficulties make vertical integration a more cost-effective and flexible option. Specifically, the costs of drawing up contingent contracts (autonomous contracting) to co-ordinate production between separate firms may be significantly high relative to the costs of allocating decisions internally to works management (an employment relation).

The technological interdependence argument is usually illustrated with reference to traditional 'flow-process' industries — for example, iron and steel industries, and brewing. It clearly applies, however, to production methods in industries where, especially in the post-war period, process innovations are increasingly continuous technologies (Nasbeth and Ray, 1974). In this regard, we should note Davies's comments below, particularly for their relevance to understanding the organisational (and spatial) logic of 'just-in-time' production. According to Davies 1987: 88):

Technological interdependence is probably more important as a determinant of the degree of integration built into the plant or firm from its outset, than as a motive for changes say through vertical merger. In established technologies, interdependencies will be well understood and recognised as new plants are built; yet *where technologies are undergoing radical changes, new interdependencies might arise*, thus encouraging vertical merger or internal expansion into adjacent stages.'

Williamson's (1986c) criticisms of other 'production function' models of vertical integration extend to Stigler's (1951) elaboration of

Adam Smith's theorem that 'the division of labour is limited by the extent of the market'. We also saw, at the end of Chapter Three that Teece (1980) invokes transaction cost arguments to debunk the notion that economies of scope necessarily result in multiproduct firms. According to Stigler's 'life-cycle' hypothesis, where the firm comprises technologically separable production processes subject to different cost behaviour, the organisational form of individual industries varies over the product life-cycle. The existence of what Silver (1984) — after Williamson — calls 'information impactedness' problems initially requires firms in a new industry to develop their own machinery, labour skills, distribution channels and finance (vertical integration). This 'do-it-yourself' mode of industrial organisation gives way, with significant market expansion, to a more complex social (inter-firm) division of labour, as specialist firms emerge to exploit the scale economies available from supplying the industry as a whole (vertical disintegration). The final stage of the industry's 'life-cycle' is marked by market shrinkage, such that 'these subsidiary, auxiliary and complementary industries begin also to decline, and eventually the surviving firms must begin to reappropriate functions which are no longer carried on at a sufficient rate to support independent firms' (Stigler, 1951: 190). Most economic geographers will be familiar with the putative spatial implications of Stigler's hypothesis: localisation to achieve the early gains of inter-firm specialisation, followed by geographical dispersion only after industries have grown 'large', and, then, spatial reconcentration induced by market degeneration.

In his critique of Stigler's model, Williamson makes the (previously stated) argument that organisational form can not be explained by (direct) production efficiency considerations — in this case, scale economies rather than technological interdependencies (or scope economies). Why must the scale economies be achieved within one firm? (see also: Silver, 1984: 131–7). The answer, according to Williamson (1986c: 205–6), turns on the transactional difficulties inherent in inter-firm rivalry, which pre-empt the prospective gains from specialisation in decreasing cost activities. Where opportunism and bounded rationality behaviour prevail, it is argued, conditions of monopoly supply (arising from specialisation in a new industry with small numbers of firms) act to undermine exchange relations based on long-term or short-term contracts. My understanding of these transactional difficulties is that they arise from problems of uncertainty or imperfect information inherent in buyer–supplier relations, but which are particularly significant in the 'new industry' scenario described by Stigler. For example, in the absence of developed markets and without the benefits of 'learning-by-doing', buyers would not have the information necessary for evaluating suppliers' price and quality control specifications (Barzel, 1982). The hazards of inter-firm contracting ('transactional difficulties') would, under these market conditions (and assuming the prevalence of bounded rationality and opportunism), have a decisive influence on industrial organisation rather than production technology *per se* (scale economies).

The basic thrust of Williamson's own transaction cost analysis, as it applies to vertical (dis)integration, is suggested by my previous discussions: under what circumstances are the costs of market exchange so high or so low, such that, respectively, integration or disintegration is all the more likely? In so far as market exchange is articulated through different forms of contract — here, in intermediate product markets — Williamson's analysis is directed at identifying the basic factors which determine the relative efficiency of alternative contractual arrangements, or the optimum choice of organisational form.

His consideration of transaction cost determinants, as we have seen, focuses on the interplay of two sets of basic factors, namely: the 'human factors' of bounded rationality and opportunism ('self-interest with guile'); and, the 'transactional factors' of market uncertainty, frequency and asset specificity. The general effect of uncertainty, when coupled with bounded rationality, is to impede long-term contracting, owing to the substantial transaction costs involved in negotiating for future contingencies. Thus, under conditions of substantial market uncertainty, the more flexible and cost-effective option is internal organisation:

> With internal organisation, issues are handled as they arise rather than in an exhaustive contingent planning fashion from the outset. The resulting adaptive, sequential decision-making process is the internal organisational counterpart of short-term contracting and serves to economise on bounded rationality. [Williamson, 1986c: 201]

A series of short-term contracts may, therefore, also provide for adaptive, sequential, decision-making and for 'economising on bounded rationality'. There are, however, further transactional difficulties inherent in short-term contracting which arise from the joint effects of small-numbers relations and opportunistic behaviour. According to Williamson, large-numbers competition (at the time of the first contract bid) commonly tends to degenerate into small-numbers bargaining (at the time of contract renewal), owing to the competitive advantage which the winner of the initial contract develops through 'learning-by-doing'. Thus, at the contract renewal stage, buyers and suppliers are effectively involved in a bilateral trading relationship (rather than a competitive market), such that a protracted 'game' of contract negotiations is encouraged. The costs of bargaining, in this situation, are substantial where transactions are frequent, and there are consequently strong incentives for the firm to pre-empt costly haggling through internal organisation (use of the authority relation). Investments in computer-based monitoring and control systems provide opportunities for lowering high-frequency transaction costs, thereby making inter-firm contracting preservable. In this event, the authoritative role of these technological innovations is analogous to that of the 'electronic line-judge' in professional tennis, which reduces the haggling costs of 'close-to-the-line' calls.

The preservation of trading relations (whether or not by a 'electronic line-judge') offers real economic value due to what Williamson calls the widespread condition of 'asset specificity'. In general, non-trivial asset specificity implies that both parties to a transaction — firm–firm, employer–employee or landlord–tenant — are 'locked into' an economic relationship. I referred earlier to the condition of 'site specificity' that arises from the 'lock-in' effects of production location imperatives in flow-process industries (transacting problems are resolved by placing plants under common ownership). This same condition may apply to other forms of durable assets, as Davies (1987: 93) explains:

> For example, firm A makes a contract with firm B for the latter to provide him with a certain specific input, whereupon B buys and installs specialised capital equipment in order to produce that input. If there is no other readily available customer for that input, A may try to renegotiate terms, safe in the knowledge that his supplier B is 'locked in'. Indeed so long as B is offered enough to cover his operating costs plus a premium to make it just worthwhile to continue supplying rather than selling the equipment for scrap, in the final event he will (presumably reluctantly) agree to the

renegotiated terms. It is not necessary that such recontracting should occur, only that it is a possibility, for integration to be an attractive proposition'.

This hypothetical situation may also apply to the employment relation and transacting problems that make the internal labour market a preferable alternative to spot-contracting (Williamson, Wachter and Harris, 1975). Here, the idiosyncratic knowledge obtained by employees through 'learning-on-the-job' gives rise to small-numbers relations and opportunistic behaviour (much like inter-firm relations). Non-trivial asset specificity, in this case, refers not simply to sunk investments in specific human capital, but more broadly to information impactedness problems (the substantial costs of transferring experiential knowledge between employees).

The general implications of non-trivial asset specificity (coupled with Williamson's 'human factors') is, therefore, that both parties to a transaction have a vested interest in maintaining the continuity of trading or employment relationships. In the case of inter-firm relations, the optimum contractual arrangement designed to achieve this continuity may be vertical integration. Instead of a simple black and white option between 'markets' and 'hierarchies', however, the requisite degree of flexibility and adaptability may be more efficiently achieved through various other forms of so called 'relational contracts' and 'quasi-integration' (Klein, Crawford and Alchian, 1978; Blois, 1972; Aoki, 1984).

Williamson's analysis of vertical (dis)integration has been criticised on a number of counts. Consider, for example, McGuinness's (1987) remarks on Williamson's 'efficiency' explanation of organisational form:

> In particular, given the importance attached by Williamson to efficiency, there is no clear treatment of the trade-offs that might be necessary between short-term and long-term efficiency. One example of the inadequacies that arise from such informal presentation of the theory is the absence of a clear analysis of the interdependence between production technique and organisational form. Instead of focusing on this interdependence, as one would expect of a truly dynamic theory, Williamson discusses the efficient resolution of contractual problems that are associated with given techniques of production. The techniques themselves, it would seem, are regarded by him as exogenous to his analysis. If so, only limited claims, if

any, should be made for the theory as an explanation of the historical evolution of the firm, and of industrial organisation. [p. 58]

In addition to highlighting the joint influence of transaction costs and production costs on the 'efficient' form of organisations, McGuiness's comments draw attention to the choice of organisational structure as a strategic variable. The strategic nature of organisational structure implies, then, that simple cost-minimising is not a decisive factor in its choice. As noted by Jacquemin (1987: 139):

> Indeed, an additional organisational cost could be more than compensated for by the increase in revenue made possible by the better control of the market resulting from the adoption of this structure.

Thus, in explaining organisational form, due consideration should be given to the nature of the firm's strategic behaviour in different types of market environment. With respect to inter-firm, rivalistic, behaviour in product markets, Casson (1987: 14) states that:

> If a firm, for example, faces the same duopolistic rival in many different markets for its product then interdependence between marketing decisions will be far greater than if the firm faces many quite separate rivals in each market. This suggests that, so far as the geographical decentralisation of marketing decisions is concerned, a unitary configuration, which centralises market decisions, will more often be preferred to a multi-divison configuration when rivalry stems from a single duopolist than when it does not.

Market power or monopoly considerations extend, of course, to the labour market. Here the choice of organisational form may be central to achieving greater bargaining power over trade union organisations (Cowling, 1984) and higher degrees of managerial control in the labour process (Putterman, 1984). In the same vein as Holland's (1976) earlier statement on 'Capital Versus the Regions', Casson comments:

> The MNE [multinational enterprise] must recruit unionised labour, and this can lead to a bilateral monopoly situation in the locality. Few unions are organised on an international basis, however, and so it is difficult to pursue a coordinated international bargaining strategy against the MNE. The MNE can therefore play off a labour union in one locality against a

labour union in another locality without fearing that the two unions will collude by refusing to accept work that has been switched from one locality to another. [p. 19]

A rigorous analysis of the nature of the firm's market environment — in so far as it has some bearing on the choice of organisational form — would ultimately need to take account of relevant government policies and, more broadly, the social environment within which firms operate. For example, in evaluating Williamson's 'Market and Hierarchies' (1975) approach with respect to small-firm growth in the Modena area of northern Italy, Lazerson (1988) concludes that:

Before firms choose markets or hierarchies, they are affected by state policies that promote certain forms of business combinations more than others, often for reasons that are socially and politically, rather than economically motivated. The strategies of Italian small firms would appear very different if tomorrow the entire legal structure promoting artisanal firms was withdrawn. This is also true for the special labour laws that make some organisational choices more attractive than others. [p. 340]

The basic conclusion to be drawn from this section's discussion is, therefore, that equal consideration should be given to 'efficiency' and 'strategic behaviour' arguments in analysing processes of vertical (dis)integration, and changes in market 'environments', more generally. Further, we should note that, with respect to operationalising Williamson's framework, transaction costs are notoriously difficult to specify, not least to measure. The 'measurement problem' has the obvious implication that the importance of transaction costs — relative to other motives for changing organisational form — is difficult to establish (see, however, Flaherty, 1981; Monterverde and Teece, 1982). More broadly, the problem of specification is remarked upon by Jacquemin (1987) as follows:

Establishing a particular form, such as the M-form, that is superior to all other forms, such as a long-term contract, a buying and stocking joint venture, a quasi-integration, or an association of the Zaibatsu type, requires such a restrictive specification of transaction costs that its formulation is implausible ... trying to determine optimal modes of organisation simply on the basis of production and transaction costs is utopian. [p. 149]

Information-intensive production

We saw, from the case studies in Chapter Five, that manufacturing firms are major users of computer network technology. With the diffusion of process innovations such as 'just-in-time' systems, based

Table 6.1 *Functional role of information exchange in manufacturing*

Originating function	Information	Utilising functions
Design	Product structure	Development Process planning Master production schedule planning Materials planning Manufacturing Purchasing Technical publications Sales
Process planning	Routing sheet	Master production schedule planning Capacity planning Manufacturing
Purchasing	Open purchase orders	Materials planning Capacity planning Manufacturing
Master production schedule planning	Master production schedule	Purchasing Manufacturing Sales
Sales	Customer orders	Master production schedule planning Materials planning Capacity planning Manufacturing Purchasing
Manufacturing	Operation details	Materials planning Capacity planning Sales

Source: A. Weatherall (1988), *Computer Integrated Manufacturing*, Seven Oakes, Butterworths, p. 91.

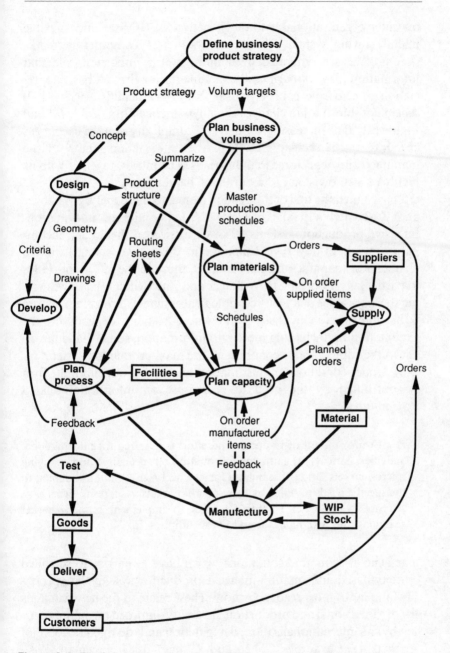

Figure 6.1 *The information flows in a hypothetical manufacturing firm*

Source: Weatherall (1988), p. 92

on inter-organisational computer networks (IONs), and 'flexible manufacturing systems', based on local area computer networks (LANs), we are reminded even more of the important role that information plays *throughout* the manufacturing firm. A basic appreciation of this role can be gained from Weatherall's (1988: 91–2) schema of information changes for a hypothetical firm (Table 6.1 and Figure 6.1). The increasing volume and complexity of these exchanges account for the secular growth of the 'information sector' of the manufacturing workforce in industrialised countries, together with its highly refined division of labour (see Chapter Two).

I have taken the title of this section from a recent paper by Willinger and Zuscovitch (1988), 'Towards the economics of information-intensive production systems'. The discussion in this paper encompasses much of my own thinking on qualitative changes in the character of manufacturing production, through the 1970s and 1980s in particular. According to Willinger and Zuscovitch, we are witnessing the emergence of a new 'technological system', whose main feature is its 'information intensity', and whose viability depends on the exploitation of 'increasing returns to information, somewhat equivalent to the capital-scale economies in the mass production system' (p. 245). Under conditions of information-intensive production, the firm is confronted by the need to economise on information-related expenditures:

> A growing amount of resources and effort is required for informational activities, namely: (a) gathering information on specialised and changing micro-markets; and (b) searching for new technical solutions adapted to the specific needs revealed by these micro-markets. As a result firms have to cope with higher transaction costs to implement refined market searching and with higher R and D expenditures. [p. 246]

The general economic conditions which have given rise to so-called 'variety-based, information-intensive production systems' are probably familiar to most readers by now. These relate to the new demands for organisational and production 'flexibility' imposed on manufacturing by the internationalisation, deregulation and 'de-massification' of markets. (See example, Cohen and Zysman, 1987; Freeman, 1987; Piore and Sabel, 1984; Roobeek, 1987). As stated by Cohen and Zysman, the slogan of the 'new era' of manufacturing is 'flexibility', applied not only to changes in production technique but also to changes in industrial organisation:

> A new era in manufacturing organised around programmable automation seems to be beginning. Its symbol is the robot. Its mythology is being built around stories such as that of the machinery-component producer who made frying pans when demand was slack for its primary product. One central concept is flexibility. Other notions that will congeal as a new orthodoxy will likely prove to be vertical disintegration and reorganisation. [p. 152]

Recent geographical literature is seemingly replete with the 'flexible' properties of manufacturing firms and 'new' industrial districts (see, for a review of the literature, Gertler, 1988). Before turning to this body of literature, let us briefly consider 'flexibility' in the context of our present discussion of 'information-intensive production systems' (IIPS).

The behavioural property of 'flexibility', in current parlance, derives from higher degrees of information intensity throughout the manufacturing firm. In factory operations, flexible manufacturing systems (FMS; based on industrial LANs) that permit economies of scope to be realised in non-standardised, 'batch', production (programmable machinery/output variety/short runs) involve processes of factor substitution marked by a growing information intensity in the types of capital and labour deployed. This is differently put in Aglietta's (1979) analysis of the transformation from 'Fordist' to 'neo-Fordist' labour processes, and in Warskett's analysis of changing factor combinations using Porat's (1978) 'revised' model of the traditional neo-classical production function (information capital, information labour; non-information capital, non-information labour).

The growing information intensity of direct production work, attendant to FMS innovations, derives from basic changes in the 'architecture' of a labour process reorganised to achieve economies of scope *both* in fixed (but 'flexible') capital and fixed (but 'flexible') manpower resources (Graham and Rosenthal, 1986). One major result is labour-savings (smaller workforce) and capital-savings (smaller plant). Another potential outcome of these technical changes in capital–labour relations, at the level of factory production, is what Aglietta calls the rise of the 'polyvalent worker', created out of a recomposition of skills previously scattered across the 'Fordist' assembly line, and consequent pressures on firms to 'redesign' jobs and implement more 'flexible' work practices (Elger, 1987). According to Ebel (1985), then:

> Generally speaking, productive activities decline and production support-ing activities increase. Also the requirements for manual skills decrease and some become obsolete, while cognitive skills gain in significance. [p. 139]

The growing information intensity of productive operations derives also from a new form of relational exchange between manufacturing firms, commonly referred to as 'just-in-time' (JIT) systems. Implemen-tation of such systems has the basic objective of developing sustained competitive advantage in end-user markets through supply chain partnership. Reductions in inventory costs throughout the 'chain' are permitted by shorter manufacturing lead-times and frequent through-put of small quantities. This form of exchange relationship, in so far as it requires high levels of integration between the principal firm and its suppliers across a wide range of functions (e.g. purchasing, engineer-ing, production, marketing and materials management), necessitates close inter-firm co-ordination in product development, quality assu-rance and logistics (Frazier, Spekman and O'Neal, 1988). As such, the use of JIT systems — the substitution of information for inventory — contributes to the growing information intensity of manufacturing operations:

> JIT exchanges necessitate frequent communication between the firms involving a large number of participants across many functional areas. Transaction costs [i.e. the costs of running a relationship; Williamson, 1975] are likely to be moderate to moderately high throughout a JIT relationship, especially in the beginning as the supplier and OEM [original equipment manufacturer] adapt to a new way of conducting business. [Frazier et al., 1988: 54–5]

The substantial transaction-specific assets committed to JIT exchange relationships — at the stages of negotiation, implementation and monitoring–enforcement — account for their long-term contractual nature, or description as a form of 'quasi-integration' (Antonelli, 1988b). They are basically information-intensive production linkages, involving also the transfer of technological and managerial 'know-how' within manufacturing industry, which confers competitive advantage on both the principal firm and its supplier. It should, however, be noted that the counterweight to production flexibility at the level of individual contracting firms is organisational 'inflexibility' at the level of inter-firm relations. As such, our understanding of JIT systems and their diffusion should be based (again) not only on a consideration of production and organisational 'efficiency' factors but also on analysis of strategic behaviour and inter-firm power relations within and across different market environments (see, for example, Pfeffer and Salancik, 1978; Robins, 1987).

The viability of flexible automation', whether organised within or across manufacturing firms, depends quite logically on increased resource commitments to the key informational functions of R and D and marketing. In the former case, the high fixed costs of 'continuing innovation' (truncated product-cycles) tend to encourage organisa-tional flexibility, in the shape of collaborative R and D projects between firms and between firms and government (including higher education institutions) (Howells, 1989; Charles, 1988), as well as internal reorganisation based on integrating the work activity of specialists — production engineers, designers, marketing managers, scientists and so on (Camagni, 1988). This latter form of 'functional integration' is discussed by Camagni as follows:

> In fact, the objective of a speedy introduction of a competitive product on the market is increasingly reached by forcing engineers, applied researchers and marketing personnel to work together in 'mission' units from the conception phase to industrialisation and production, on an efficient (and not just a prototype) scale. This is particularly valid in the new technology industries, where the product life-cycle is shorter and the need for continuous innovation stringent. [p. 54]

The general effect of collaborative R and D projects and 'functional integration' is to raise transaction costs and, thereby, the firm's information expenditures. The scope for organisational flexibility, in this case, is clearly limited by the adjustment costs (a capital cost) involved in a major transformation of the firm's internal and external communication channels (Arrow, 1984: 170–1). According to Malmgren (1961), for example:

The information strategies and decision rules adopted over time are partly dependent upon information in previous periods, and are conditioned by it, with the result that the rate of adjustment to new kinds of activities is dependent upon the universality, or lack of it, in the informational structure of previous periods. Besides, the rate of accumulation of new information from research activities is limited by the costs the firm is willing to incur and the time strategy adopted. More can be had faster, but more people have to be employed to work it out. [pp. 419–20]

The growing information requirements of manufacturing are also met by the buying-in of variegated producer services — for example, information relating to industrial engineering, process design and research, as well as to market environments (MacPherson, 1988). The increasing centrality of the manufacturing firm's marketing function — the third leg of the 'three-legged stool', with the others being R and D and production — derives again from the intensification of competition brought on by internationalisation, 'de-massification' and deregulation trends. These developments, acting to fragment demand and to heighten uncertainty, necessitate higher levels of expenditure on product differentiation and on the acquisition of market knowledge — that is, relatively more of the firm's resources are committed to market research, sales, advertising, etc. A major result of this intensification of market intelligence-gathering required to sustain variety-based FMS is higher levels of integration between manufacturing firms and the distributive sector (Marshall, 1988; Milne, 1989).

With the growing information intensity of manufacturing, Willinger and Zuscovitch (1988) argue, the competitive position of the firm turns on its ability to exploit 'informational economies of scale' (see Wilson, 1975; Arrow, 1984). The basic methods required to achieve these types of scale economies are the codification and standardisation of information exchanges, and the strategic co-ordination of the firm's external network of contractors and potential collaborators (see, for example, Benson, 1975; Thorelli, 1986). In this emerging context, information and the new technology are central to the future development of manufacturing industry:

The growing integration of information technologies shifts the economic emphasis from capital-intensive to information-intensive production. If scale economies diminish in importance, inputs and outputs variety increase and transaction costs rise. To overcome transaction costs induced by the continuous re-definition of micro-markets and rising R and D costs, such 'Information-Intensive Production Systems' should prove their economic viability by finding some means of increasing efficiency. We have argued that, indeed, new technologies provide such a potential. Statically, the economies of scope of flexible equipment may compensate for information costs. Dynamically, a process of 'algorithmisation', that is, codification and routinisation of decision-making via more and more information-processing, may increase organisational efficiency, widen the strategic scope of the firm, and 'internalise' more control over the environment. [Willinger and Zuscovitch, 1987: 253]

Very clearly, this section's discussion provides a highly simplified view of the growing information intensity of manufacturing and its underlying causes and effects. I hope, nevertheless, that the bare essentials of this perspective on changes in the character of manufacturing firms are sufficiently clear, such that the following observations on geographical research are seen to have a rationale.

Information linkages and spatial agglomeration

The previous discussion suggests that information linkages are playing a more important role in the locational dynamics of manufacturing industry. Indeed, this is evinced by recent literature on the 'new' industrial geography (Storper, 1987), and consequently I have proclaimed the 'comeback' of information space (Hepworth, 1989). In elaborating further on this 'comeback', I will focus attention on what seemed to be new and different about information space, but this circumscription of available literature is done for sake of simplicity rather than to eclipse or invalidate our rich inheritance of knowledge on the economic geography of information.

The 'new' aspects of information space I wish to concentrate on are twofold. First, and related to the discussion in Chapter Five, there are information linkages which lie at the centre of 'functional integration' in manufacturing firms, and whose spatial ramifications are extended by technological innovations such as 'flexible manufacturing systems' (FMS) and 'just-in-time' systems (JITS). Second, and related to this chapter's first two sections, there are information linkages spawned by processes of 'vertical disintegration' in manufacturing, where transaction costs are thought to exert a critical influence on the spatial agglomeration of industry. In the latter case, particularly, some geographers appear to be asking a more complex question than that posed by Thorngren (1970) nearly two decades ago, namely: How do contact *and* contract systems affect regional development? Let us now explore these 'new' facets of information space, illustrating our discussion with a few individual contributions to the literature. My immediate purpose is to highlight only what this recent geographical work may reveal about the locational behaviour of manufacturing activity in the information economy.

The analytical focus of Camagni's (1988) research on Italian manufacturing is 'functional integration' within large firms, induced by competitive pressures for higher rates of product and process innovation. Here, the creation of information linkages within the firm — to interweave 'conception' (research, design, engineering) with 'execution' (direct production) tasks — is the formal method of functional integration, with the new information technologies providing operational efficiency (see also Ciborra, 1983). My understanding of Camagni's analysis is that these organisational and technological innovations favour metropolitan areas as industrial locations for three main reasons. First, the greater information intensity *and* spatial 'compactedness' of the firm's production process combine to increase and localise its demands for advanced producer services and highly skilled labour, which tend to be concentrated in metropolitan areas. Second, and given this 'locational pull', the operational efficiency of 'flexible' production systems requires strong information linkages and spatial proximity with the firm's manufacturing suppliers, particularly where JIT innovations are in use (Estall, 1985; Lambooy, 1986; Meyer, 1986). And third, close collaboration between management and labour organisations is needed to facilitate new patterns of work organisation, and, at least, in the early stages of the innovation process these transaction costs are incurred centrally:

> The demand for nearby contacts is influenced by the frequency of conflicts and negotiation conditions and, when it is possible to standardise information exchange, nearby contacts can be substituted by telecommunication networks. [Johansson, 1987: 56]

The locational implications of functional integration, noted by Camagni, are partially confirmed by Schoenberger's (1987) study of technological and organisational innovation in the American car industry. With respect to spatial agglomeration tendencies induced by computerised, 'flexible', production systems, for example, she points out that:

> The flexibility that is required by the intensification of competition and permitted by the new forms of automation also requires considerable flexibility on the part of suppliers of parts and sub-assemblies. If the final assemblers are competing on the basis of a high level of product differentiation, they must be prepared to act swiftly to the changing composition of final demand. This, as noted, is one of the key attributes of the new automated system: they permit a shifting mix of product configurations on the line without sacrificing scale economies. Frequent modification of products on the assembly line also requires a very rapid response of suppliers as their product configurations and product mix change as well. The smooth functioning of the flexible line would seem to benefit strongly from spatial proximity of suppliers and final assemblers. [p. 204]

Importantly, and unlike Camagni, Schoenberger's discussion of the possible spatial implications of functional integration and its attendant technological innovations is extremely circumspect. She surmises, for example, that a more spatial extensive separation of the production process may be induced by the firm's market geography. Indeed, the case studies presented in Chapter Five strongly underscore this 'possibility' (see also Milne, 1989). Further, where production segments are standardised, a corporate strategy of 'dispersed concentration' may be motivated by the perceived risk of trade union militance in older industrial areas. In this last respect, and as noted briefly in Chapter Five, the 'flexible' capabilities of computer network technology offer management opportunities for enhancing their control over the labour process. According to Aglietta (1979: 127), for example:

> Automatic production control divides up the production processes in a new way, linked on the one hand to a centralisation that permits direct control of production by long-distance transmission of information, and on the other hand to a rearrangement of the segments of a complex labour process to allow for important savings in transfer time, quality control and the preparation of production programmes. A far more advanced centralisation of production becomes compatible with a geographical decentralisation of the operative units (manufacture and assembly).

It is clear, so far, that information linkages are emerging as key factors in the location dynamics of manufacturing industry. What is not clear, however, is the overall spatial direction towards which these strengthening 'linkages' are tending. This is hardly surprising, at present, given the complexity and embryonic nature of the technological and organisational innovations we are attempting to comprehend. Let us now complicate matters even further by considering a second explanatory basis for understanding the rising importance of information linkages within manufacturing industry and their attendant spatial effects.

Attention is focused here on the recent work of Scott (1983, 1986b), particularly his application of transaction-cost economics to study recent change in the functional and spatial organisation of manufacturing industry. This work is noteworthy, in my view, for its attempt to provide an intermediate level theoretical apparatus for use by economic geographers, whose interests lie in 'mapping out' the putative decomposition of Fordist mass production relations. Thus, while Scott's analysis is immersed in the turbulent context of 'post-Fordist' and 'neo-Fordist' capitalism adumbrated, respectively, by Piore and Sabel (1984) and Aglietta (1979), its basic thrust is towards specifying a more tractable theoretical basis for exploring these envisaged transformations of the industrial landscape (Scott, 1988; Harvey and Scott, 1988). For this purpose, Scott (1983, 1986b) turns to transaction-cost economics, owing to its relevant concerns with the institutional arrangements of capitalism and with alternative forms of industrial organisation in particular (Williamson, 1985).

My specific interest is in Scott's ideas on the spatial implications of vertical disintegration in manufacturing industry. It is these ideas, after all, which lie at the heart of Scott's and others formulation of regional development through 'flexible specialisation' (see also Storper and Christopherson, 1987). According to Scott, then:

> As firms disintegrate vertically (and horizontally too for that matter) so the level of external transactional activity in the economy increases. This, in turn, stimulates those producers with especially intense and costly linkages to one another to converge locationally around their own centres of gravity. The greater the spatially determinate costs of linkage, the greater this tendency to converge will be. [1986b: 224]

The locational clustering of vertically disintegrated industries, in

Scott's view, derives from the continual need of firms to renegotiate production *contracts* and 'rebuild linkage *contacts*' (1986b). The operation of these 'transactional forces', making for the increased centrality of information linkages in spatial agglomeration, is argued to be of considerable significance under conditions of heightened market uncertainty (the 'post-Fordist world' of markets), and where transactions between firms are non-standardised and occur frequently. Importantly, as locational clusters emerge in this fashion, a process of cumulative development takes hold, with further 'rounds' of vertical disintegration being induced by the agglomeration economies usually associated with 'transactions-rich' or 'information-rich' metropolitan environments (Pred, 1977; Gottmann, 1983). As stated by Scott (1986b: 224):

> The locational clustering of many producers helps significantly to hold down the spatial costs of external transactions. This has two important side effects. First, it encourages yet further vertical disintegration by reducing search and recontracting costs. Second, and as a corollary, it encourages some vertical disintegration even among producers whose input demands are quite unstandardised and require much face-to-face intermediation (recurrent contract costs are greatly diminished where spatial proximity is secured). Thus, vertical disintegration encourages agglomeration, and agglomeration encourages vertical disintegration

Most readers following the progress of Scott's recent work will be familiar with the historical and contemporary examples of localised industrial complexes he invokes as illustrative of 'an intense, transactions-rich process of growth and development' (1986b: 225). These include not only the much-celebrated 'Third Italy' (Brusco, 1982) and the 'Hollywood' film industry (Storper and Christopherson, 1987) but also metropolitan producer services (Daniels, (1986) and high-technology 'Sunbelt' complexes (Scott and Storper, 1987). The present significance of these 'transaction-rich' industrial complexes, according to Scott (1988), derives from their potential role as 'harbingers' of a new capitalist mode of regional development based on 'flexible production systems'.

My interpretation of Scott's argumentation is that if research in industrial geography is to progress — to encompass the spatial dynamics of 'post-Fordist' or 'neo-Fordist' production systems — we must develop a rigorous theoretical understanding of the process of vertical disintegration (and integration) in manufacturing firms. As Storper and Christopherson (1987) emphasise, 'the vertical disintegration that lies behind flexible specialisation creates powerful agglomeration tendencies at the regional level, such that the spread of this form of industrial organisation threatens to bring about significant changes in existing patterns of industrial location, inter-urban trade, and inter-regional growth transmission' (p. 115).

What emerges strongly from Scott's analytical treatment of vertical disintegration, and from Storper and Christopherson's study of 'flexible specialisation' in the US motion picture industry, is the central role played by information linkages in spatial agglomeration. The importance of these linkages derives, of course, from the agglomeration forces associated with transaction cost-minimising behaviour at the level of the manufacturing firm. In turn, the necessity of focusing attention on the role of transaction costs in industrial location stems from our interest in the changing boundary which separates 'the firm' from 'the market' (vertical integration/disintegration) and its attendant spatial ramifications, whether this falls under the rubric of an economic geography of 'post-Fordism', 'neo-Fordism' or not.

Conclusion

It is undoubtedly the case that both information and the new information technology are currently playing a key role in the spatial restructuring of manufacturing industry. However, and at the risk of becoming monotonous, it is clear that we are still at the stage of agenda-setting as a basis for refining and broadening the thrust of geographical research. The existing literature on 'flexible production' (of whatever variant) is replete with 'caveats' (Schoenberger, 1987) and warnings for the reader to avoid 'attaching any sort of teleological meaning to these processes' (Scott, 1983: 227). The 'new' industrial geography is, then, an intellectual area where 'angels fear to tread' — but 'tread' they must!

In concluding, there are at least two research issues which I believe warrant more fundamental investigation. First, we need to develop a more critical view of the role of transaction costs in vertical (dis)integration, given their presumed strong influence on the locational dynamics of 'flexible production'. As emphasised in this chapter, the transaction cost 'framework' goes no more than halfway towards explaining why vertical (dis)integration occurs, and, indeed, its static efficiency properties hardly enables us to model regional development as a dynamic process of economic and social change. The analytical power of the 'framework' depends, then, on supplementary consideration of the strategic motivation and behaviour of firms and amongst firms, and equally, on due consideration of changing technological, market and political environments. In this last respect, we must take information technology seriously, given the powerful capabilities of these innovations for altering the transaction-cost structure of firms and markets — that is, the cost basis of vertical (dis)integration and 'flexible specialisation'. Also, it is incumbent on researchers to specify the local and inter-regional expression of environmental uncertaintly that is attributed to the 'post-Fordist transition', if it is this 'uncertainty factor' which currently operates dramatically to change the social division of labour in regional economies.

Second, in specifying the process of technological change itself, there is a clear need to take account of the full range of computer network applications, given the pervasive role of information linkages within the manufacturing firm (see Figure 6.1). In my view, industrial geographers have been highly selective in focusing on flexible manufacturing systems and 'just-in-time' systems (reflecting perhaps their proclivity with direct production), to the extent that the functional integration capabilities and locational 'flexibility' afforded by computer network applications as a whole are untowardly neglected or 'left hanging'. The empirical work on computer network innovations which I have conducted (Chapter Five), of course, has the reverse bias, highlighting even more the need for a more comprehensive specification of technological change.

In the next chapter, I turn to the subject of 'informatics in capital markets', and specifically to the role played by information and the new technology in the global integration of these markets. This 'sector-hopping', from manufacturing to financial services, is not without reason. Not only must the adoption of 'flexible production systems' be (partly) financed through resort to external capital markets but also the internationalisation of these 'systems' has been underpinned by new and seemingly even more 'flexible' arrangements for securing the global movement of financial capital.

Chapter Seven

Informatics and Capital Market Integration at a World Scale

Introduction

The backdrop to the last chapter, the *raison d'être* of flexible production, was the enveloping presence of global economic uncertainty, which heightened some two decades ago. Greater flexibility in production, whether achieved through technological innovations (e.g. flexible manufacturing systems) or changes in industrial organisation (e.g. 'flexible specialisation'), are not the only responses to significantly higher levels of economic uncertainty or risk. A further, important, option for firms is, at least in the short term, to alter their capital structure in favour of the most flexible of all forms of investment — financial assets. This other dimension to the 'flexible firm' is neatly articulated by Arrow (1984: 160) as follows:

> When there is uncertainty, risk aversion implies that steps will be taken to reduce risks. This partly affects decisions within the firm, such as the holding of inventories and preference for flexible capital equipment, and partly leads to new markets which will shift risks to those most willing and able to bear them, particularly through the equity market.

In the recent spate of geographical literature on so-called 'flexible accumulation', I have encountered few attempts (if any) to explore the interrelationship between production and financial flexibility, despite its widely recognised importance to regional adjustment (Clark, Gertler and Whiteman, 1986) and to spatially determinate processes of industrial restructuring more generally (Bluestone and Harrison, 1982). In particular, the 'financial services revolution' appears to have drifted over the heads of industrial geographers, to be consigned to that 'catch-all' sector we refer to as 'producer services' (Thrift and

Leyshon, 1988; Leyshon, Thrift and Daniels, 1987). Ironically, for Scott and Storper (1987), the producer services sector epitomises 'flexible specialisation' in a metropolitan context.

This chapter is, then, about financial rather than production flexibility, and it focuses attention on the global restructuring and integration of capital markets. Once again, I suggest that current developments in the information economy are at the centre of structural changes in industries and markets, with information technology playing an identical role in product diversification (the creation of new financial instruments), geographical diversification ('24-hour' equities trading) and in market regulation (the 'information agreements' which collectivise transaction costs on the 'electronic trading floor').

Information in capital markets

The economic justification for security markets, as stated by Ohlson (1987:101), 'stems from the existence of uncertainty and the need to allocate attendant risks; information identifies the perceived uncertainties, and therefore it may also affect the choice of a consumption pattern'. What complicates the microanalysis of information in capital markets (and in other markets throughout the economy) is the real-world presence of so called 'imperfections'. The latter may, for example, emanate from 'trading noise' (French and Roll, 1986), such that market prices imperfectly convey information among dealers; further distortions to the 'perfect' functioning of capital markets may arise from differential time-lags in the publication and reception of information (Cukierman, 1984).

In general, these informational 'imperfections' in capital markets, including the market for equities, are the rule rather than the exception (Grossman, 1976). Importantly, it is these very 'imperfections', acting to undermine complete certainty in market decisions, which provide traders with an incentive to acquire information — that is, to incur considerable expenditure on information about securities. Thus, in the case of 'time-lags', Cukierman notes (1984: 33):

> When there is a lag, however small, between transactions and the latest observed price, there is always an incentive to collect costly information

because non-informed traders can use the information available in current market prices only in the next trading period. As a result an informed trader manages to retain an advantage during the current trading period.

Time-delays in disseminating price information between trading centres, and their shortening by successive improvements in communication technology (telegraph, telephone, 'ticker tape', etc.), are central to historical explanations of the integration of national and international securities markets (Garbade and Silber, 1978). The theoretical inspiration for these studies of 'boundary change' in securities markets is generally traced to Stigler's (1961) classic paper on 'The Economics of Information'. According to Stigler, in large markets, the costs of search are so great that 'there is a powerful inducement to localise transactions as a device for identifying potential buyers and sellers' (p. 66). The growth of single financial centres, in most countries, is likewise explained by Kindleberger (1983) in terms of the economies of scale associated with centralisation — that is, the reduction of transactions costs, especially those of search.

Stigler also suggests that the significant enlargement of markets will bring forth a set of firms which specialise in collecting and selling information:

> They may take the form of trade journals or specialised brokers. Since the cost of collection of information is (approximately) independent of its use (although the cost of dissemination is not), there is a strong tendency toward monopoly in the provision of information: in general, there will be a 'standard' source for trade information. [p. 73]

Indeed, today, the world's security markets are underpinned by an immense information industry, but in which the information monopoly of national stock exchanges is rapidly being eroded by multinational corporations offering on-line market-making services — for example, Reuters and Telerate. I will return to these competitive developments below. However, even with free entry into this burgeoning information industry, Stiglitz notes (1982: 119) there will remain 'partial monopolies in which particular individuals may have a monopoly of particular pieces of information'.

In examining the nature of the incentives for information acquisition in the stock market, therefore, financial economists tend to distinguish between 'private' and 'public' information. The former type includes much of the information produced by investors and security analysts, and it affects prices through the process of market trading; public information, on the other hand, affects prices before anyone can trade on it, and examples include the financial reports of firms and governments. Information used by investors falls in the continuum between public and private information (French and Roll, 1986).

As shown by Stiglitz (1982), the problems posed by asymmetrical or differential information in capital markets can be recast as problems in the economics of screening. A special case in point relates to the information provided by firms — for example, annual reports — which is certified by 'public' market-making institutions, such as the stock exchanges. In so far as the retention ratio[1] of the firm can convey important information about the productivity of the firm — to competitors, governments, trade unions, as well as investors — there is an incentive for firms to provide 'misinformation' or partial information in their public reporting of financial results. Similarly, in so far as the market value of the firm (one of the determinants of productive investment decisions) is sensitive to trading knowledge of its actions, incomplete information may have implications for not only the level of investment but also for the choice of (less observable) techniques. According to Stiglitz (1982: 153):

> If scale of investment conveys information, then either unproductive firms will over-invest in order to obfuscate the difference between their firms and the more productive firms, or the more productive firms will over-invest in order to differentiate themselves from the less productive firms.

The disclosure laws, or 'information agreements' (O'Brien and Swann, 1968), pertaining to the market-listing of firms on stock exchanges are intended to minimise the production of private information. Established regulatory bodies, such as the Securities Exchange Commission in the US, similarly attempt to reduce the amount of private information produced — for example, insider trading laws attempt to control profits earnt by corporate officials on their information (Jaffe and Rubenstein, 1987). Nevertheless, the economic problems of 'screening', as they interact with incomplete diversification out of risk among individual investors due to incomplete information, are endemic to capital markets. For example, in the present context of international portfolio diversification (see below), investor preferences have favoured 'blue chip' corporations whose brand-name products and ubiquitous production activities are relatively familiar to institutional and individual investors in different countries (Yang, Wansley and Lane, 1985). As such, information asymmetries in the 'global' equities market have disproportionately benefited the largest multinational corporations, such as Imperial Chemical Industries (ICI) and International Business Machines (IBM).

Thus, it is clear that the economic implications of imperfect information in capital markets are several fold. They range from their effects on patterns of productive and financial investment to the induced growth of specialised information industries. By corollary, technological innovations in financial markets and markets for financial information must have a significant impact on the organisational structure of the capital market and its attendant geography, as well as indirectly influencing the general direction of investment activity in the rest of the economy. However, before turning to the new information technologies' role in capital markets, it is more logical to pre-consider the financial innovations and market restructuring processes they have made feasible.

Financial innovations in world markets

Although innovations in new financial instruments have a long history, the last few decades stand out as a period of 'revolutionary' change in the world monetary system (Kindleberger, 1987). Indeed, the dramatic changes presently underway in capital markets encouraged Cooper (1984) to predict, for the year 2010, the formation of a single world monetary authority with responsibility for overseeing a fixed set of world exchange rates and a single world monetary policy. Similarly, in advancing his vision of a new financial world without barriers, Walter Wriston, the ex-chairman of Citicorp (1970–84), declared:

It is fair to say that [scientific advances in communication] have created an entirely new system of world finance based on the incredibly rapid flow of information round the world. I would argue that what one might call the information standard has replaced the gold standard and indeed even the system invented at Breton Woods'.[2]

The suspension of the convertibility of the dollar into gold, announced by the US government in August 1971, was basically a national policy response to the prevailing international monetary crisis which the general shift to floating exchange rates was expected to resolve. This shift, with its subsequent effects of greater international mobility of capital, was intended to 'permit automatic correction to the post-war chronic balance of payments deficits that plagued the major trading nations' (Stonham (1987: 2). The general, and more fundamental, effect of these exchange-rate adjustments was, however, the emergence of 'a more permanent new view of international risk and return' (Stonham 1987). Financial innovations during the 1970s and 1980s reflected basically this 'new view', and were stimulated by the changing structure of economic incentives; for example, the period's significant increase in interest-rate volatility reduced the attractiveness of fixed-rate debt instruments and conventional securities flotations (Silber, 1983).

The demand–supply framework, used by the Bank for International Settlements (BIS) (1986), is a useful and relatively simple basis for analysing the development of these numerous financial innovations. The latter are classified by the BIS according to the type of financial intermediation performed, that is: risk transfer, liquidity enhancement, credit generation and equity generation (Table 7.1). I will look at changes in the equity market in the following section.

The most important class of this 'new wave' of financial innovations consists of instruments and services that permit economic agents to transfer asset price risk. The enormous growth in demand for these types of innovations is attributable to greater volatility in real exchange rates and real interest rates, and, more fundamentally, to changes in their surrounding policy regimes — respectively, the widespread adoption of floating exchange rates and 'floating' interest rates (the switch to quantitative monetary aggregates as intermediate policy targets). Innovations in credit-risk transfer techniques, according to the BIS, were accelerated by the collapse of the energy (oil) sector boom and the Third World debt crisis as they converged around mid-1982. In turn, the last of these developments induced liquidity-enhancing innovations, as the credit-worthiness of international banks was called into question. This heightened concern over banks' financial exposure, acting together with the demand for more liquid transactions holdings (the effect of higher and more variable interest

Table 7.1 *A classification of innovations by financial intermediation function*

Innovation	Price-risk-transferring	Credit-risk-transferring	Liquidity-enhancing	Credit-generating	Equity-generating
A. On-balance-sheet					
Adjustable rate mortgages	x				
Floating rate loans	x				
Back-to-back loans	x				
Asset sales without recourse		x	x		
Loan swaps		x			
Securitised assets		x	x		
Transferable loan contracts		x			
Sweep accounts and other cash management techniques			x		
Negotiable money-market instruments			x		
Money-market mutual funds			x		
Zero coupon bonds				x	
'Junk' bonds				x	
Equity participation financing				x	
Mandatory convertible debentures					x
B. Off-balance-sheet					
Futures	x				
Options and loan caps	x				
Swaps	x				
Forward rate agreements	x				
Letters of credit	x	x	x		
Note issuance facilities		x	x	x	
Credit-enhancing guarantees on securities		x		x	

Source: Bank for International Settlements.

rates), caused a shift in investor preference from traditional deposit accounts to capital-market instruments. Increased leveraging, particularly higher corporate debt equity ratios, also fuelled the demand for liquidity enhancements.

Undoubtedly, the dominant force behind credit-generating innovations was the vortical effects on global investment flows of US financial deficits. The US reliance on foreign purchases to sustain its debt-funding is described by Hamilton (1986: 114–15) as follows:

the biggest base on which many of the securities and investment houses have thrived, has been the enormous and growing size of the US deficits. The accompanying ballooning of the market for US Treasury bonds and the regular new supplies of bonds put up for auction have ensured that American government debt has become the key reference point of interest rates all around the world and, perversely, the greatest supporter of the dollar ... Japan has at various times in the last few years bought as much as a third of new Treasury bill issues and the proportion of total US debt that Japanese institutions held at the end of 1985 had reached well over 20 per cent. Another 10 per cent was held by the Swiss, the West Germans and other global investors.

A major effect of the US Treasury's debt-financed fiscal expansion was to 'crowd out' corporations from domestic capital markets. For new sources of credit, to finance domestic mergers and acquisitions as well as foreign direct investments, US corporations looked increasingly to the Eurobond market (see Table 7.2), which now supported large-scale, primary share, issues.

The major *suppliers* of these financial innovations have been the world's largest banks and securities houses. In the wake of stock market deregulation in the US, marked by the end of fixed broking

Table 7.2 *The rise of the Eurobond market ($ million)*

	1963	1970	1980	1983	1985
Number of issues	13	128	310	526	1 357
Value	147.5	2762.0	26 423.0	46 376.5	135 676.2

Source: Euromoney.

commissions in 1975, there was a wave of diversification and acquisitions among 'Wall Street' institutions out of which emerged powerful new financial conglomerates. These hybrid organisations 'were able to use their greater capitalisation to capture high-volume securities business on finer margins by different sorts of corporate alignment that did not violate the (Glass–Steagall) Act' (Stonham 1987: 5) — the Act separating commercial and investment banking in the US. A direct result of this Act's operation was to encourage the internationalisation of the American financial services industry, as noted by Greenspan (1988: 94):

> In the corporate debt market, for example, US banks' foreign subsidiaries served lead roles in underwritings approaching $17 billion in 1986, or about 10 per cent of the volume of such debt managed by the 50 firms most active in the Eurosecurities market last year. These and other essentially investment banking activities have permitted banks to continue to service those customers seeking to rely increasingly on securities markets — provided that the securities are issued abroad.

The general effect of 'deregulation' in US capital markets, and the ensuing development of new financial products (e.g. options and futures) and new computer-based trading systems, was to export the 'financial services revolution' to all other countries (Hamilton, 1986). For example, the so called 'Big Bang' in the London stock market (October 1986) — the ending of dual-capacity brokerage and fixed commissions, and the entry of foreign firms as members of the Stock Exchange — parallels competing 'liberalisation' measures in other European financial centres and in the still highly protected capital market of Japan (the *Economist* Publications, 1986; Coakley and Harris, 1983). What has emerged, therefore is a global market for new financial services characterised by an intense struggle for competitive and comparative advantage among the world's leading multinational finance corporations and leading financial centres.

Although the originating turbulence of the world monetary system may subside in the future — particularly with multinational initiatives such as the European Monetary Agreement (Zis, 1984; Sarcinelli, 1986) — the BIS predicts the continued development of financial innovations. The main stimuli of further innovation are, according to the BIS, provided by the global integration of economic structures and the institutionalisation of the process of financial innovation within the firm (see also Lessard, 1988).

Capital market integration: a 'slippery concept'

The financial innovations described above have contributed to the development of a 'global' capital market transcending national economies. In practice, capital markets lie somewhere between full integration and full segmentation. National markets are not genuinely integrated in the sense that a significant number of savers do not distinguish among borrowers on the basis of nationality (Cooper, 1972). Also, given that interest rates are generally manipulated by governments in the pursuit of domestic economic objectives, the 'law-of-one-price' notion of capital-market integration is hardly realistic — the same real interest rate is not offered in all of the world's stock markets.

In a world of nationally differentiated taxes, capital controls, political risks and so on, empirical studies of capital-market integration are made problematic by what is really 'an operationally slippery concept' (Kohlhagen, 1983). Nevertheless, the much-publicised emergence of 'round-the-clock' trading enabled by global, inter-market flows of information and money, the growth of foreign direct investment in financial services and 'deregulation' trends have all led to a general acceptance that the world's capital markets are becoming increasingly integrated (Hepworth, 1988a). The 'law-of-one-price' may not, indeed, operate, but there is strong circumstantial evidence to support this basic proposition – as the worldwide equity market 'crash' of October 1987 clearly underlined (Dobilas, 1988; Bank of England, 1988). Let us, then, briefly consider some aspects of this process of global market integration, focusing on equities trading.

Some small economies, such as Sweden and the former British colonies of South Africa and Australia, have a relatively long history of raising equity capital in foreign markets — liquidity for financing economic expansion was simply unavailable at home. The last decade has, however, witnessed a more general trend towards the multinational placement and trading of equities, stimulated not only by the search for liquidity but also by investor strategies directed at reducing risk through portfolio diversification across national borders. As with the Eurobond market, the rapid growth in international equities trading has been enabled by a succession of policy 'liberalisation' measures — for example, the abolition of withholding tax and exchange controls, the relaxation of controls on foreign membership of national stock exchanges, the harmonisation of market regulations that affect corporate listing and the generalised erosion of non-tariff barriers which have, historically, protected domestic capital markets.

The most common argument for international investing is provided by modern portfolio theory: given that the world's equity markets do not move together like 'synchronised swimmers', there are opportunities with equity portfolios to reduce significantly risks for a given return, or to increase the return with a given level of risk by diversifying across stock markets as well as within them (Bergstrom, Koeneman and Siegel, 1983). Indeed, the relatively late interest shown by US financial institutions in foreign equity, particularly pension funds, is partly explained by national differences in the diversification potential of investors' own stock market. As stated by Rutterford (1983: 312):

> For example, the risk of the German stock market has been found to be 44% of the average risk of holding one German share, whereas the comparable figure for the US is 27%. US market risk is lower because the US stock market is larger and offers greater diversification opportunities than its German or UK counterpart. Thus, the German investor would benefit more than the US investor from international diversification. This is one reason why European investors in general have a greater international content in their investment portfolios and why US investors have only recently begun to take an interest in the opportunities offered by investment in foreign currency securities.

A further implication of size differences in the world's capital markets (see Table 7.3) is suggested by the extension of the capital asset pricing model (CAPM) to an international context.[3] If all of these capital markets were strictly 'integrated', the model's application would lead to the conclusion that the US should account for the largest investment in any individual equity portfolio — that is, an optimal 'world market portfolio' would include countries according to the size of their stock markets rather than their low correlation coefficients (Rutterford, 1983: 313–15). In reality, of course, investors do not distribute their country investments according to this model for several reasons, such

Table 7.3 *The size of the world's stock markets, 1985: equity market value and turnover, £ million*

	Market Value	Share (%)	Turnover	No. Companies Quoted Domestic	Foreign	Foreign Share (%)
1 New York	1 302 241	42.3	671 280	1 487	54	3.5
2 Tokyo	648 674	21.1	271 481	1 476	21	1.4
3 London	244 711	7.9	52 777	2 116	500	19.1
4 Germany (Association of Exchanges)	123 783	4.0	68 825	451	177	28.2
5 Toronto	108 867	3.5	21 823	912	54	5.6
6 Zurich	76 161	2.5	n.a.	131	184	58.4
7 Basle	69 151	2.2	n.a.	320	523	62.0
8 Paris	58 477	1.9	14 522	489	189	27.9
9 Geneva	55 061	1.8	n.a.	271	537	66.5
10 America	43 740	1.4	18 214	732	51	6.5
11 Australia (Association of Exchanges)	41 611	1.3	22 455	1 069	26	2.4
12 Amsterdam	41 200	1.3	14 181	232	242	51.0
13 Hong Kong (All Exchanges)	40 792	1.3	n.a.	247	22	8.2
14 Milan	39 377	1.3	10 885	147	0	0
15 Johannesburg	37 579	1.2	1 671	501	26	4.9
16 Singapore	27 918	0.9	2 480	122	194	61.4
17 Stockholm	25 820	0.8	7 578	164	7	4.1
18 Brussels	14 014	0.4	2 482	192	144	42.8
19 Madrid	13 730	0.4	2 248	334	0	0
20 Barcelona	12 298	0.4	377	458	0	0
21 Kuala Lumpur	11 181	0.3	1 754	223	61	21.5
22 Copenhagen	10 451	0.3	224	243	6	2.4
23 Luxembourg	9 106	0.3	55	n.a.	n.a.	n.a.
24 Oslo	7 060	0.2	2 914	156	7	4.1
25 Tel-Aviv	5 290	0.2	408	267	1	0.4
26 Kuwait	4 716	0.2	190	38	4	9.5
27 Helsinki	4 244	0.1	209	50	1	2.0
28 Vienna	3 182	0.1	567	64	38	37.2
29 Athens	527	–	11	114	0	0
(TOTAL) (3,080,562) (100.0)						

Source: The Stock Exchange, London.

as the differential transactions costs, taxes and information between domestic and overseas investment.

The CAPM's prescriptions are, nevertheless, made more note-worthy by the accelerated erosion of these cost differentials. For example, the recent development of composite market indicators — so-called 'World Financial Indices' — convey real-time information on price changes to investors who exhibit a strong, growing, interest in world market portfolios — namely, global investment trusts, pension funds and international banks. In addition to this easing of the informational constraints on international diversification, foreign securities have been made more attractive by the types of financial innovations discussed earlier. For example, the effects of exchange-rate volatility, which Eun and Resnick (1988) suggest accounted for about 50 per cent of the volatility of dollar returns from investment in major overseas stock markets between 1980 and 1985, can now be mitigated by risk-reduction financial innovations, such as, multicurrency diversification and hedging via forward exchange contracts.

A further important aspect of the growth of transborder equities trading is the development of the Eurosecurities market.[4] The latter has grown particularly with the wider acceptance of international depository receipts (American/European), which enable companies to issue equity on most of the world's major stock markets relatively quickly and without incurring the cost of multiple listing (Stonham, 1987). Basically, the recent expansion of this form of 'quasi-equity' trading, which derives from sales techniques used in the Eurobond market, is part-and-parcel of a more general trend towards 'securitisation', or the shift from debt to equity financing in capital markets. Other examples of these new types of financial instruments, for hedging against or 'unbundling' risk, are stock index options and futures.

The internationalisation of stock markets has also encourged a growing number of multinational corporations to obtain listings on several national exchanges (Dobilas, 1988). As stated earlier, the major beneficiaries/users of this more direct route for raising equity capital are the most well-known of the international corporations:

> The names involved are predictable so far: large companies like ICI, BAT and Glaxo from the UK; Honda, Hitachi, and Matsushita from Japan, Royal Dutch Shell and Phillips from the Netherlands. In other words, global 24-hour markets may only be made in a few big names for quite a long time to come. This is underpinned by the natural conservatism of investors. The second limitation arises from the inherent difficulty of analysing stocks on a transnational basis, with problems of different accounting standards and disclosure. [Stonham 1987: 13]

Some recent data, for the US, illustrate the size and growth of foreign securities issues and transactions in corporate debt (see Tables 7.4 and

Table 7.4 *Foreign transactions in US corporate securities, 1985–87 ($ millions)*

	1985	1986	1987 (Jan–Oct)
STOCKS:			
Foreign purchases	81 995	148 101	221 545
Foreign sales	77 054	129 382	195 611
BONDS*:			
Foreign purchases	86 587	123 149	93 135
Foreign sales	42 455	72 499	67 263

* Includes state and local government securities, and securities of US government agencies and corporations. Also includes issues of new debt securities sold abroad by US corporations organised to finance direct investments abroad.

Table 7.5 *New security issues by US corporations, 1984–87 ($ millions)*

BONDS only:	1985	1986	1987
Public issues, domestic	74 175	119 559	231 936
Private placement, domestic	36 324	46 195	80 761
Sold abroad	22 613	37 781	42 596

Source: Federal Reserve Bulletin, Volume 74, No. 2., February 1988, Appendix of Tables, Nos 34 and 65.

7.5). The general effect of these corporate financial strategies is, of course, to integrate international capital markets. First, the (share) holders of these corporations' liabilities are offered diversification gains from the broader spread of investment across different national stock markets. And, second, by using several stock markets to finance direct investments in different countries, multinational corporations gain from the arbitrage opportunities offered by international differences in the average costs of capital. The effect of these capital arbitrage activities is, then, to integrate stock markets at the global level (Caves, 1982). As a result, however, financial economies of scope will diminish:

This increasing integration of financial markets implies, of course, an evening of the cost of funds in various countries and a consequent reduction in the benefits accruing to a firm from spanning national financial markets. At the same time, however, global competition puts more pressure on multinational corporations to take advantage of the remaining gains from global financial scope. [Lessard, 1988: 5]

While today's capital markets provide multinational corporations with greater flexibility in their financial activities, the 'price' of this flexibility has correspondingly increased owing to the greater informational requirements of controlling geographically dispersed operations. Innovations in computer and advanced telecommunications technology have been crucial to the achievement of these global control imperatives — for example, in the area of international fund management. More generally, the types of financial innovation described earlier, and the process of international capital-market integration itself, could not have emerged over the last decades in the absence of these powerful information technologies. Their enabling role in capital-market restructuring is, therefore, worth highlighting.

The role of information technology

New information technology has played a multifaceted role in the global process of capital-market restructuring. The application of computer network systems has facilitated the development of new equity-related products with complex financial structures, such as options and futures, while supporting the more complex trading strategies employing these new types of financial instrument. This has accelerated the growth of not only cross-market trading of related financial products but also inter-market trading of the same securities (Scribner, 1986). The internationalisation of this trading activity has, in particular, depended on the '24-hour' use of network technology for monitoring and controlling financial exposure at a global level.

A major force behind the internationalisation of equities trading has been the emergence of new computer-assisted markets, such as NASDAQ and Instinet,[5] which operate as worldwide communication systems rather than being confined to a central physical location. The expansion of NASDAQ, which now ranks third only to the New York and London stock exchanges in terms of total market capitalisation (see Table 7.3), was accelerated in the US by the 1975 Securities Act Amendments, which called for the creation of a National Market System based on the exploitation of electronic market linkages. The rationale of this legislation was that the application of new technology would enhance market efficiency and competition through the improved exchange of information between dealers. Whereas central-ised exchange trading gave specialist dealers direct and immediate access to trading information, with 'remote' (say, relative to New York) dealers being forced to follow the market by telephone and daily quotation sheets, the basic thrust of NASDAQ was to eliminate this spatially biased market structure through the decentralised provision of information. The general effect of these changes in market-place organisation, according to Hamilton (1978), has been to increase marketability of common stocks in the US (including those of foreign companies), while extending the geographical scope of the equities market.

The more recent internationalisation of NASDAQ is exemplified by its planned interconnection with the London Stock Exchange's SEAQ (Stock Exchange Automated Quotations) system. The latter, which currently operates as a price-dissemination service for several hundred securities traded internationally, has developed out of the Exchange's more established TOPIC network (Figure 7.1). What is expected to emerge from the NASDAQ–SEAQ link, is a transatlantic equities trading facility, complete with computer-based price quotation, surveillance and other market-place functions.

The NASDAQ–SEAQ link-up, which parallels several other global, inter-exchange connections (Hepworth, 1988a), is basically a defensive market strategy induced by intensifying competition from third-party public information providers — in particular, Quotron, AP-Dow Jones and Reuters. Until recently, the first two of these public data base vendors held a temporary global information monopoly over the joint distribution of current price and other market data on trading in Japanese (Tokyo/Osaka) and US stocks. In March 1987, Reuters purchased the rights to distribute SEAQ services which give buy-and-sell quotations in UK equities from the London Stock Exchange's market makers. Additional to these data purchases, Reuter's global business strategy has involved the acquision of other smaller computer services firms, including I.P. Sharp, which offered a ready-made

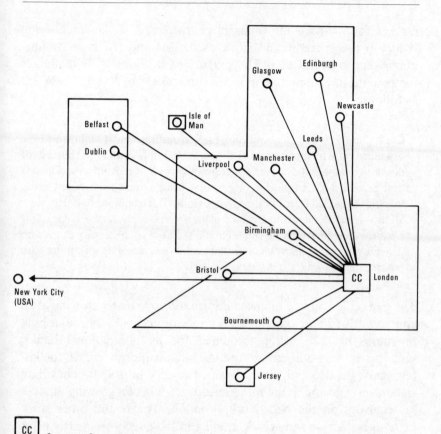

| CC | Computer Centre (incl. on-line data bases) |

O   Multiplexer/s supporting 12 Topic terminals each in local sub-network

———— Telecommunications link

Figure 7.1 The London Stock Exchange TOPIC network

Source: Hepworth (1987a)

packet-switching network (see Figure 5.5) for the creation of an international 'electronic exchange' infrastructure. The company's main competitor, Telerate, has pursued a similar path of vertical integration, and, in December 1987, announced its intention to create a joint 50:50 stock company with AT and T to develop a dedicated trading system (Murphy, 1988).

As such, we are witnessing growing transnational competition in the market for financial information, characterised by an intensifying rivalry between traditional stock exchanges and the new 'on-line' public data base industry. The general implications of these market forces in the information economy is summarised by Knight (1984: 15) as follows:

> Capital markets are all about information and the dealer will turn to the organisation which can provide the information he wants. In the days of domestic trading and before the technological era, that information was best provided by people meeting in one place at particular times, using facilities provided by a stock exchange floor. The telephone and telex now allow international trading to take place without a physical presence. It is clear that the international capital markets lie not so much with those who provide a physical floor where trading can take place, but with those who control the information systems.

The market for on-line financial information, currently growing at an annual rate of 30 per cent, is highly concentrated, with the major bulk purchases of data being accounted for by international banks, securities trading houses and multinational corporations.[6] In the last few years, the demand for worldwide financial information has risen significantly, owing to the simultaneous trading of a growing number of securities on the New York, London, Tokyo and other stock exchanges. 'Around-the-clock' trading in these securities partly takes the form of passing 'book' positions among branch offices in several different countries — and between different time-zones.

In order to monitor and control this global spread of trading activity, financial institutions are investing heavily in transnational computer networks for the timely movement of variegated information between foreign branches, and between dispersed branches and corporate head offices (Bruce, Cunard and Director, 1987). For example, Prudential–Bache Securities, the New York-based stock brokerage and investment bank, operates satellite communication links to its regional offices in London and Hong Kong; Morgan Stanley's international network provides the company's capital markets group with real-time access to both the London and Tokyo trading arena; leading Japanese institutions, such as Nomura Securities, are the driving force behind major telecommunications developments such as the Tokyo and Osaka teleports and the trans-pacific fibre optic cable, with the former being directly 'aimed' at the US mid-west and west coast financial centres.[7] One overall effect of these organisational technological innovations has, quite obviously, been the internationalisation of the financial services industry, through foreign direct investment (Hepworth, 1988a). The extent of this geographical diversification, in the cases of London and New York, has been studied, for example, by Leyshon, Thrift and Daniels (1987).

Importantly, the role of the new information technology also extends to other key functions in the equities market. These include the computerisation of the order routing and executing process, the clearing and settlement process, and the confirmation and accounting process. As noted by Minnerop and Stoll (1986), these types of 'back office' or trading-support services are increasingly contracted out to specialist firms, who incur the necessary capital investment in computers and software to provide these services to brokerage houses (Daniels, (1988). Similarly, the specialised electronic funds transfer services of the SWIFT (Society of Worldwide Interbank Financial Telecommunications) have played a critical enabling role in the internationalisation of securities trading, and the financial services industry in general (Langdale, 1985).

Perhaps less obviously, technological innovations in capital markets are being used for surveillance. An example is the American Stock Exchange's on-line Stock and Options Watch system, which provides regulators with enhanced capabilities in the detection of possible uses of non-public information, market manipulation and market-maker misconduct (Scribner, 1986). Unlike trading activity itself, these surveillance applications of computer network technology do not have a global reach — that is, regulatory bodies, such as the Securities Exchange Commission (SEC) in the US, neither have the information nor the jurisdiction to monitor 'off-shore' market processes which can potentially affect domestic securities markets. This spatially differen-tiated bias in the technology's control and regulatory applications, therefore, has provided dealers with additional incentives to interna-tionalise their trading activities.

We can, therefore, see that the new information technology has played and continues to play, a central, multifaceted role in the global restructuring of capital markets. The internationalisation of the equities market, for example, could simply not have occurred on its present scale without the application of computer networks to trading and basic market-place functions; the development of new derivative, equity-related products (options, futures, etc.), even more so, could not have occurred in the technology's absence. At the same time, these organisational, technological and financial innovations have all acted to highlight the role of information in capital markets, from its regulatory function to its more direct function in portfolio diversification at a global scale.

Emerging research issues

Quite clearly, the types of capital-market developments I have outlined pose an immense challenge for geographical research, let alone economic policy-making at different spatial scales. This is due to the great complexity of world capital markets and the myriad of direct and indirect influences they exert on the rest of the economy. It would, therefore, be unrealistic to offer a completely worked-out research agenda on the subject of 'world capital markets and geography'; instead, what follows are tentative suggestions for future research that follow some recent lines of inquiry.

First, there are research issues pertaining to the consequences of international portfolio diversification for regional development. These apply to country 'regions' within the global economic system and to regions within individual countries, and they relate particularly to the investment strategies both of private and public financial institutions. As we saw earlier, as capital markets are becoming increasingly integrated at the global level, investors have greater flexibility in the geographical scope of their financial strategies. The possible implications of these strategies are suggested by Kindleberger (1987: 84):

> Savings would be gathered in the world financial network, largely from richer countries, but also from rich people in the poorer ones, and distributed to those countries (or rather those businesses in all countries) where the marginal productivity of capital was high, whether because of local scarcity of capital due to poverty and inadequate savings, or because of rapid growth and surging innovation.

The logic of Kindleberger's argument is now to be found in recent empirical studies of inter-regional patterns of portfolio investment in individual countries. For example, in their geographical assessment of selected pension fund investments controlled by the State of Wisconsin Investment Board, Botts and Patterson (1987) found that 80 per cent of recent stock purchases (made by the Board) was directed primarily to the so-called 'Sunbelt' states — thus, calling into question 'the degree to which SWIB is accomplishing its acknowledged "invest in Wisconsin" policy' (p. 25). An identical line of research inquiry has been pursued by Minns (1982) and Coakley and Harris (1983) with respect to the foreign orientation of the UK's pension funds, whose portfolio diversification strategies, it is argued, have contributed to structural decline in domestic manufactuing industry — a failure to 'invest in the UK'. As Botts and Patterson rightly emphasise, however, we would need to know how pension funds based in other parts of the US allocate their own portfolio investments geographically in order to offer a balanced assessment of these capital flows affect regional development. Clearly, in the context of global capital-market integration, I would add that we will increasingly need to account for the international geography of (say) the Wisconsin Investment Board's portfolio diversification and the US state's absorption of non-US equity investment (see Table 7.4).

The vulnerability of Third World countries in this New International Financial System (NIFS) has been highlighted by Thrift and Leyshon (1988). A second set of research issues, raised particularly in this 'Third World' context, concerns differential access to the all-important market for financial information. As emphasised by Helleiner (1983), at a time when the complexity and variability of conditions in international capital markets have underlined the role of information or knowledge in investment decisions, developing countries are threatening to be locked-out of the NIFS owing to its surrounding, informational market barriers. As stated by Helleiner (1983: 14–15):

The provision of information has become a lucrative business for large investment and merchant banks, the services they provide ostensibly kept entirely separate from their interest in lending. Among the best known advisory services are those provided by the 'triad' (Kuhn Loeb Lehman Bros International, New York, Lazard Freres and Co., Paris and S. G. Warburg, London), although many other banks have recently set up similar services. The question for developing countries is whether it would not be in their interests to establish alternative means of expertise and advice, from their own information consortium and enable better control over advisers.

In fact, the general significance of uneven development in the global market for financial information has also been recognised by the industrialised European countries. In the European Community, where more than 50 per cent of this market is accounted for by US public data base vendors,[8] policy concerns extend to the future development of information industries in general — including tele-communications services and computer manufacturing. Further, US pressure for including financial services in GATT trade agreements, together with US dominance of the financial information industry, is seen by European governments as a considerable threat to the full integration of the Community's internal capital market.

Related to these trade issues is a third set of research concerns, which have to do with the availability of information for economic policy control. As central trading institutions, such as the national stock exchanges, have lost their information monopoly in capital markets, the information needed for macroeconomic regulation has become correspondingly more dispersed. For example, the value of statistics on international capital flows has been impaired by the shift from bank intermediation to bond finance; similarly, as banks have resorted more to multiple-market (rather than close-customer) relationships where price is the main signalling device, the quality of lenders' information about ultimate borrowers' credit-worthiness may have worsened (Dini, 1984). The basic thrust of policy debates here is that a more distributed information structure in capital markets, under conditions of 'bandwagon' investor speculation and computer-assisted trading, can contribute to greater short-term market volatility (Bank for International Settlements, 1986). And, as stated by Kindleberger (1987: 84), 'To the extent all market participants are tuned in to the same information and forecasts on a world scale, and closely observe each other's actions, wide changes of opinion become possible and with them large international movements of funds.'

These important policy-related issues, therefore, relate to the difficulties of controlling economies raised by the informational effects of capital market integration (Langdale, 1985). More broadly, as financial integration has proceeded, the respective importance of the interest rate and the exchange rate in monetary policy has changed. The general effect of this change, which will have differential implications for monetary policy in relatively 'open' (e.g. Canada) and 'closed' (e.g. the US) economies, is that developments in the major capital markets are rapidly transmitted to the real domestic economy, influencing inflation, relative prices and the level and composition of output (Kincaid, 1988).

A fourth set of research issues relates to what Andreff (1984) calls the 'international centralisation of capital'. These centralisation tendencies, applying to the linked internationalisation of productive and bank capital, are particularly evident in the securities markets. As stated earlier, the main beneficiaries of the so called 'world equity market' are the best known, 'blue chip' multinational corporations, whose total cost of borrowing is reduced by shopping around in several different markets, including the Eurosecurities market. The multinational listing of shares may have other significant implications for foreign direct investment patterns in the global economy. For, as Caves (1982: 93) points out: 'An alternative to the joint venture is for the multinational enterprise to issue minority shareholdings in a subsidiary on the host country's securities market.'

These 'blue chip' companies have, at the same time, provided a platform for the foreign market penetration of large securities firms and large banks who have depended on 'blue chip' shares in the risky process of international portfolio diversification (Coakley and Harris, 1983). Thus, as Cooper (1972: 239) suggests, 'the growing international capital market may foster the concentration of industry'. Further, these trends towards concentration are being actively encouraged by national regulatory authorities, whose lifting of restrictions on ownership linkages between the banking and securities industries is generally motivated by concerns over the international competitiveness of their respective countries' financial institutions (OECD, 1987).

Although multinational corporations are more able to exploit the capital arbitrage opportunites offered by a 'global equities market' (relative to firms operating in single countries), the realisation of these financial scope economies requires production and organisational flexibility which is both costly and difficult to sustain (Baldwin, 1986). In the case of managing operating exposure, multinationals may configure individual businesses to have 'the flexibility to increase production and sourcing in countries that become low-cost producers due to swings in exchange rates', or, 'to reduce operating exposure by matching costs and revenues' on a country-by-country basis (Lessard, 1988). The second option implies that financial considerations will lead to a divergence from the optimal geographical configuration of the firm's productive operations, in terms of scale and locational advantages. A further option in managing the exchange-rate uncertainty (translated into earnings and cash flows) inherent in multiple sourcing of capital markets is to shift the incidence of risk to other firms or local suppliers. This flexibility in the multinational's business strategy may be achieved in the area of finance — say, the choice of invoicing currencies — and/or in the area of production — say, contracting out of component manufacturing.

All of these four research areas present an exciting but difficult challenge to economic geographers. They clearly underline the need to go beyond the financial services sector itself in considering the geographical implications of capital-market restructuring at a global scale. While some recent studies have shed light on the localised effects of these restructuring processes on selected 'world city' financial centres (for example, Leyshon, Thrift and Daniels 1987), our current state of knowledge on their general significance for urban, regional and national development is extremely poor.

Conclusion

The most dramatic and visible impacts of the information economy have occurred in the world's capital markets. The advanced usage of the new technologies by financial services firms is explained by the intrinsic information intensiveness of their products and production processes. Importantly, the recent 'boom' in capital markets can be attributed to the shift from fixed to floating exchange rates in the early 1970s, which generated demand for new and more flexible forms of financial instruments (and fixed capital) to counteract higher levels of global risk and uncertainty. Against this turbulent background, the 'information standard' came to replace the 'gold standard', and, with this conversion, the role of information technology has become crucial to the operation of the world's capital markets, the profitability of financial conglomerates and, more indirectly, to the investment and locational strategies of the markets' dominant users — multinational corporations.

The several research issues I have identified merit considerable attention. Of particular interest is how world capital-market restructuring is altering the geography of information and risk, and thereby inter-regional flows of financial capital used for productive investment. Similarly, what role does financial flexibility play in productive and managerial decentralisation within multinational corporations (Sabel, 1989; Young, Hood and Dunlop, 1988)? Clearly, it will be increasingly necessary for geographers to become more 'internationally minded', as well as less 'sectorally minded'. In this last respect, financial flexibility matters for industrial geographers; and equally production flexibility matters for the geography of services.

The spatial and economic implications of global capital-market integration have been primarily studied by geographers with an interest in producer services. Attention is focused here on the impact of internationalisation on the local economies of the world's leading financial centres, such as London and New York. Some researchers, such as Moss (1986a), have highlighted the creation of new technological infrastructure in 'world cities', which has arisen directly out of the information and communication requirements of the 'financial services revolution'. This new urban infrastructure comprises public and private computer networks. In the next chapter, I examine some dimensions of this emerging 'information city' — the urban nexus of the information economy.

Notes

1. The fraction of its own shares held by the firm can convey significant information to its competitors, with respect to productivity and overall corporate performance.
2. Cited by A. Hamilton (1986), *The Financial Revolution*, Harmondsworth: Viking/Penguin, p. 30.
3. For the model's application to international portfolio diversification, see J. Rutterford, *Introduction to Stock Exchange Investment*, London: Macmillan, 1983, Chapters 8 and 10.
4. Eurosecurities are securities placed outside their country of origin. For example, in 1984, abut 15 per cent of British Telecom shares (£4 billion) were placed outside the UK.

5. NASDAQ (National Association of Securities Dealers Automated Quotations System) was introduced in the early 1970s, subsequently revolutionising over-the-counter market trading. Instinet, conceived as a global electronic market for trading in both listed and over-the-counter stocks, is now owned by Reuters.
6. See Information Dynamics Ltd, report on *Financial Information Services in Europe*, prepared for the Directorate Generale for Information Market and Innovation, Commission of the European Communities, London, September 1985.
7. See L. Kotlowitz, 'Making it Means Spending it', *Communications Week*, 3 August 1987, Close-up supplement, c. 4.
8. See Information Dynamics Ltd. op. cit., note 6.

Chapter Eight

The Information City: Further Sketches

Introduction

The strong metropolitan bias of the information economy in industrialised countries has been subject to research by numerous geographers (see, for earlier and recent literature surveys, Abler, 1974; Kellerman, 1984). Writing about the 'post-industrial megalopolis', some twenty-five years ago, Gottman (1961) observed that cities were becoming increasingly specialised in 'managerial and artistic functions, government, education, research and the brokerage of all kinds of goods, services and securities' (p. 774). More lately, he has invoked the concept of 'transactional city' to describe this pattern of economic specialisation, stating that

> In the modern world, with its expanding and multiplying network of relations, and a snowballing mass of bits of information produced and exchanged along these networks, information services are fast becoming an essential component, indeed the foundation stone of transactional decision-making and of urban centrality. [Gottman, 1983: 23]

Reflecting this metropolitan bias of the information economy, the bulk of research on the spatial impacts of telecommunications has tended to focus on structural change in urban form (e.g. Brotchie, Newton, Hall and Nijkamp [eds], 1985) and/or in city systems over time (e.g. Pred, 1977). In my view, and with some notable exceptions (e.g. Kutay, 1986a, b, 1988; Corey, 1987), geographers have made little progress in coming to terms with the major research and policy issues raised by the diffusion of information technology in cities. Much of the existing literature on this topic is woefully out-of-date and anticipatory (e.g. Berry, 1970; Abler, 1975; Goddard, 1980).

The purpose of this chapter is, therefore, to describe the nature and significance of some current developments in the metropolitan information economy. It continues my project of sketching out the 'information city' — that is, the information economy elaborated in an urban context. I regard these 'sketches', whether based on a 'neo-Fordist' model of urbanisation (Robins and Hepworth, 1988a) or simple description of technological innovation in cities (Hepworth, 1987c), as a preliminary attempt to achieve some understanding of what is clearly a complex process of societal and spatial development.

The particular 'sketches' of the information city contained in this chapter focus on two popular concepts which feature in some academic and policy debates on the urban impact of computer and telecommunications innovations: the 'electronic highways'; and the 'wired city'. I provide a critical appreciation of these concepts, drawing attention to their practical significance and adumbration as part of the economic ideology of the so-called 'information age' (Slack and Fejes [eds], 1987). Additionally, I draw on some of my current research on the role of local governments in the information economy, with particular reference to the fundamental processes of institutional change, currently under way in the UK. This discussion highlights the multifaceted nature of 'information as a social relation' (Robins and Webster, 1980).

Metropolitan cities: information-rich environments

As stated above, metropolitan cities provide opportunities for economising on the social costs of producing, processing and distributing information of various kinds (Meier, 1962). These spatial externalities apply to the operation of factor markets and final markets for goods and services. As a result of these spatial externalities in transactions, and their agglomerative force, the development of the information economy has a strong metropolitan bias (Corey, 1982). The degree of this spatial bias is evident from data on the occupational and industrial distribution of the information labour force in Greater London and Metropolitan Toronto for 1981 (Tables 8.1 and 8.2). These data illustrate not only the information intensiveness of metropolitan economies but also their highly refined division of information labour.

The seminal work of Pred (1973, 1977) focused on the historical role

Table 8.1 *The occupational composition of the information labour force: Toronto census metropolitan area and Canada, 1981*

OCCUPATION GROUP	TORONTO CMA per cent	CANADA per cent
Information Producers		
Scientific & Technical	4.1	4.2
Market Search & Co-ordination	8.8	10.3
Information Gatherers	2.6	3.0
Consultative Services	13.2	11.1
Health-Related Consultative Services	1.3	1.4
Information Processors		
Managers & Administrators	11.4	9.4
Foreman Supervisors	9.8	11.3
Clerical & Related	33.7	33.6
Information Distributors		
Educators	7.0	9.0
Public Information Disseminators	0.3	0.4
Communication Workers	1.1	0.8
Information Infrastructure		
Information Machine Workers	5.0	3.3
Postal & Telecommunication	1.8	2.0

Source: Hepworth and Dobilas (1985).

Table 8.2 *Share of information occupations in the labour force by sector: Greater London and Great Britain, 1981*

INDUSTRY SECTOR*	GREATER LONDON per cent	GREAT BRITAIN per cent
Agriculture	13.2	6.7
Extractive & Transformative	47.8	36.8
Distributive Services	61.5	53.3
Retail Services	48.9	42.9
Non-Profit Services	51.7	46.4
Producer Services	86.7	84.5
Mainly Consumer Services	45.2	36.6
Public Administration	69.1	53.0

* Industries are classified by the taxonomy suggested by J. Singelmann, *From Agriculture to Services*, Sage, Beverley Hills, CA.

Source: Hepworth, Green and Gillespie (1987).

of information and telecommunications innovations in city system development. The particular emphasis of Pred's US studies of the cumulative inter- and intra-metropolitan bias of information flows is on the locational behaviour of private and public bureaucracies. According to Pred (1977: 177):

> Large metropolitan complexes offer headquarters units and other high-level administrative activities three specialised information advantages that are seldom available to the same degree in less populous metropolitan areas and cities. These are in ease of inter-organisational face-to-face contacts, business-service availability, and high intermetropolitan accessibility.

The spatial behaviour of head-office activities has, of course, been a strong and persistent theme in the area of urban-economic geography. For example, Pred's research emphasis on head-office locations as 'control points' in the urban system was shared by Borchert (1978), and also by Semple, Green and Martz (1985); similarly, the role of face-to-face contact systems and telecommunications in office-location decisions has been studied by Goddard and Pye (1977).

The research agenda outlined by Pred has been considerably elaborated over the last decade. The influential work of Noyelle and Stanback (1983) on the 'new service economy', for example, has focused attention on higher levels of metropolitan specialisation in head-office functions and producer services. The growth of producer services, as a major aspect of increased specialisation in the metropolitan information economy, has generated considerable interest in several countries (see, for a review, Daniels, 1985). Of gathering interest here is the potential decentralising effects of information technology in the office sector (Dunning and Norman, 1987) and the locus of contracting out by large firms. In the last respect, some geographers anticipate that this 'externalisation' process will increasingly occur at lower levels of the corporate–urban hierarchy, rather than in head-office cities (as Pred suggests), due to the apparently more 'flexible' organisational structures and procurement strategies of the 'post-Fordist', 'late 1980s' multilocational corporation (Marshall, 1988).

Whereas Pred's (1977) primary spatial focus was the national urban system, the basic thrust of some recent work is characterised by its international orientation. Friedmann and Wolff (1982), for example have advanced the 'world city' concept as a basis for research on the international division of labour. From their definition of the 'world city' system, we can also discern the metropolitan structure of the global information economy:

> The dynamism of the world city economy results chiefly from the growth of the primary cluster of high-level business services which employs a large number of professionals — the transnational elite — and ancillary staffs of clerical personnel. The activities are those which are coming to define the chief economic functions of the world city: management, banking and finance, legal services, accounting, technical consulting, telecommunications and computing, international transportation, research, and higher education. [p. 320]

The 'world city' framework has been adopted by Moss (1987) to structure his analysis of national and international communication patterns in the US case. According to Moss, advanced telecommunications is creating 'a new urban hierarchy, in which certain cities will function as international information capitals, with the most extensive electronic infrastructure and richest opportunities for human interaction' (1987: 35). In essence, Moss's studies extend the work of other researchers (see, for example, the edited collection: Pool, 1977), whose studies of urban centralisation were oriented to national economies and the telephone, to an international economic arena increasingly dominated by multinational corporations and 'world city'-oriented transborder data flows (see also Espeli and Godo, 1986).

The world's 'information capitals', as control points of the 'global firm' and as international trade centres, are also the central concern of Langdale's (1983, 1987) work on metropolitan dominance in the Far East and Pacific Rim regions. In the Australian case, for example, Langdale (1987: 92) states:

> The Sydney–Melbourne head office [of transnational corporations] functions as an international information gateway: information from the firm's Australian operations is chanelled through the head office; it is sifted and repackaged and then sent to the overseas head office. Conversely the Sydney–Melbourne office functions to filter information coming into Australia from overseas sources.

Parallel to these internationalisation trends in the urban information economy are intra-metropolitan spatial tendencies, characterised by the suburbanisation of routine 'back office' operations from high-rent, high-wage central city locations (Moss and Dunau, 1987; Nelson, 1986). Of particular interest here is Baran's (1985) 'post-Fordist' analysis of an increasingly computerised labour process in the US insurance industry and its reorganisation around 'flexible' suburban labour markets. These patterns of spatial decentralisation have also received considerable attention in so-called 'telecommuting' or 'electronic homeworking' studies (see the edited collection, Van Rijn and Williams, 1988). The general effect of telecommuting, according to Zimmerman (1986), is to create 'electronic sweatshops' for female labour in what she calls a 'domesticated city'. She comments:

> Telecommuting promises two very different types of work experience for those at the upper and lower ends of the occupational scale: data entry clerks and secretaries will handle routine tasks under continuous computer scrutiny of their performance and hours, while professionals will have discretionary working hours and unrestricted freedom to use computers for personal tasks, such as home accounting and data base access. [p. 30]

The available evidence on telecommuting trends in advanced industrialised countries indicates that it applies to a very small but growing share of the labour force (Korte, Robinson and Steinle, 1987). For the most part, telecommuting projects are market and technical trials being carried out by firms and governments, in collaboration with telecommunication carriers. Importantly, the number of these 'experiments' is increasing throughout Europe (Mettler-Meibom, 1988) and the US (Ruttenberg, 1987), and technological-market forecasting studies project massive growth in the telecommuting workforce over the next decade, with the diffusion of interactive cable and 'home automation' technologies (Kinsman, 1988). The development of telecommuting, as argued by Robins and Hepworth (1988a, b), is essentially one aspect of the urbanisation process in information cities, and one aspect of 'flexible' working based on the 'domestication' of not only mental labour but also manual labour.

Parallel, but counter, to these spatial decentralisation trends in routine information processing functions is a consolidation of urban centrality in what Tornqvist (1974) calls 'non-programmable', specialised functions which demand direct personal contacts. In observing these spatial patterns, Castells (1985) states that

> the new technologies also enhance, simultaneously, the importance of a few places as locations of those activities that cannot easily be transformed into

[telecommunications] flows and that still require spatial contiguity, thus, reinforcing considerably the intra-urban hierarchy. In the informational city, spatial singularity and urban centrality become even more important than in the industrial-commercial city, precisely because of their unique locational requirements. [p. 18]

We saw, in Chapter Six, that several geographers have drawn attention to the key role of information-based spatial externalities as an agglomerative force in manufacturing complexes organised around 'flexible specialisation'. A further dimension of what Castells (1987) calls the 'high-tech' or 'informational' mode of economic development relates to the consolidation of metropolitan cities as 'knowledge centres'. According to Anderson (1986), the recent basis of this consolidation is the locational behaviour of what he calls 'the new systems-architecture-designed' corporation; this cybernetic type of firm is characteristic to the 'fourth, logistical revolution', which is 'associated with the growth of information processing and communication capacity as well as the growth of the knowledge base' (p. 9). The industrial geography of the 'fourth, logistical resolution', Anderson argues, will be structured around a new city hierarchy dominated by centres of higher education (much as Bell, 1973 suggests).

The same research theme is evident in Nijkamp's and Mouwen's (1987) work in a European context, which examines the role of inter-urban information networks in innovation diffusion and the locational dynamics of corporate R and D activity. A complementary, and important, line of research inquiry is being pursued by Taylor (1987), whose analytical focus is information-based 'core-periphery' relationships within single large corporations, and their conditioning of innovation diffusion processes at the level of organisations, regions and countries. According to Taylor (1987: 216):

Asymmetry and inequality have rarely been recognised in economic geography; this deficiency is most striking in linkage and information-flow studies. The geographical literature, in its preoccupation with space, has tended to see organisational interactions in entirely neutral terms. Asymmetric relations have been reduced in economics to the implicit recognition of only technical power, and the ability of such power to create efficiency. The exercise of positional power, probably better labelled political power within and between organisations, has been largely ignored.

Indeed, as emphasised in preceding chapters, the 'flexible' capabilities of computer network technology — for 'centralisation' and/or 'decentralisation' — tend to highlight an important question which is impossible to answer directly from studies of aggregate information flow structures or contact patterns, namely: what do we mean by the 'control points' of national and international urban systems? (Where does power and control reside?)

The 'electronic highways', or private roads?

Information flows between urban centres occur through several different media, including: radio and television broadcasts, newspapers and magazines, postal services, telex and telephone messages, and public and private computer networks for data, video, facsimile (and voice) communications. In addition, information is exchanged through direct inter-personal contacts, facilitated by air, rail and road travel. These inter-city media of communication, given their idiosyncratic applications, still co-exist in a complementary relationship with one another. For example, in the UK, the growth of 'electronic mail' has occurred in tandem with a record increase in the volume of ordinary mail traffic. Similarly, investments in local airport facilities and transport networks, generally, run parallel to the construction of so-called 'electronic highways' in the UK and in other European countries. In other words, the following discussion's specific focus on inter-city computer network media is not intended to suggest that relative accessibility and mobility in the information economy is solely determined by the geography of 'electronic highway' development.

Public computer networks are sometimes referred to as 'electronic highways', implying that they share the same public ownership and control as well as universal access characteristics which normally apply to road transportation. For most of this century, the ordinary telephone service did exhibit these public-good characteristics; however, over the last few years, telecommunications markets in several countries — such as the US, Japan and the UK — have been 'deregulated' or 'liberalised', and the historical principle of universal access has come under threat (Pike and Mosco, 1986; Gillespie and Hepworth, 1988). In the US, for example, the onus of maintaining universal access to telecommunications services has shifted from the federal to the state level, as Moss (1986b: 81) notes:

Federal telecommunication policy was guided initially by the principle of 'universal service' expressed in the Communications Act of 1934: 'to make available, so far as possible, to all the people of the United States a rapid, efficient, nationwide and worldwide wire and radio communication service with adequate facilities at reasonable charges'. Federal deregulation has shifted the demand for universal service to the state level. Through the political process, state governments remain tied to the traditional goals of keeping residential rates low, providing universal service in outlying areas, and maintaining an economically viable public telecommunications network.

In addition to city-based telecommunications planning (see below), these decentralisation trends in the US have spawned some interesting regional development initiatives. The State of California, for example, is considering the creation of a 'Commission of the Information Age', whose basic agenda addresses the following policy-related issues:

(a) the conditions that will promote continued development of a viable information industry in California that is based on the application of information and telecommunication systems;

(b) the public's need for information, the structure of the information economy, and the contribution of information to productivity and quality of work-life and leisure;

(c) questions of equity related to the proliferation of electronic systems of information collection, manipulation, storage and distribution,

(d) the requirements for citizens to be active participants in the information industry, as entrepreneurs, and workers, including their educational and cultural preparation;

(e) California's place in the global information economy, with a special emphasis on how the state can attain and maintain a position of competitive advantage in the global information economy.[1]

In Europe, 'Californian' thinking on the broader social, economic and political issues related to telecommunications development is highly differentiated between individual countries (Gillespie and Hepworth, 1988). The achievement of 'balanced' regional development in the information economy is, for example, a central objective in French, German, Swedish and Norwegian plans for advanced telecommunications development. In the UK, by way of contrast, domestic telecommunications regulation lacks a regional policy component, and infrastructure development in 'peripheral' areas is likely to depend on a number of major programmes being embarked on by the European Community — for example, STAR (Special Telecommunication Action for Regional Development) and RACE (R and D in Advancd Communication Technologies for Europe). (For a discussion of European telecommunications in an urban/regional context, see Gillespie (1987) and Goddard, Gillespie, Thwaites and Robinson (1986).)

The Japanese model of national telecommunications planning, based on the creation of interconnected 'Advanced Information Cities', has attracted considerable interest in other countries (Dutton, Blumler and Kraemer, 1987). Since 1983, the Japanese Ministry of Posts and Telecommunications has, for example, been implementing its 'Teletopia' project, which involves the experimental introduction of advanced telecommunications and information services in twenty model urbanised regions on a commercial basis. The general stated objective of the project is 'to assist the formation and development of a comfortable regional society, make full use of a new media and to promote the nationwide diffusion of new media using model cities as foci, and therefore promote Japan's smooth transition to an advanced information society'.[2]

It is, therefore, clear that the construction of 'electronic highways' in several countries is shaped considerably by market forces, or, as Melody (1986) points out, 'Telecommunications is becoming the centre piece of national industrial policies' (p. 80). A major result of this market domination by mainly large telecommunications service suppliers — protected by so-called 'natural monopoly' considerations — is that the geography of infrastructure development tends to be biased in favour of the dominant metropolitan centres of the information economy. Some evidence of this spatial bias, as it is reflected in inter-city route pricing and service availability, has emerged from European and Canadian studies of telecommunications accessibility (Gillespie, Goddard, Robinson, Smith and Thwaites, 1984). In the US case, for example, Moss (1987) argues that

Telecommunications regulation is gradually leading to the elimination of cross-subsidies and to a new telecommunications infrastructure, characterised by a multiplicity of telecommunications networks, where the choice and type of service will vary substantially depending upon the size and scale of the market. Most important, the spread of fibre optic systems, 'smart buildings' and national paging systems is already revealing that the new emerging telecommunications infrastructure will favour those metropolitan areas with information intensive industries and lead to disparities between urban and rural telecommunications systems. [pp. 12–13]

In addition to the gravitational pull of metropolitan information economies, the geography of advanced telecommunications development has been substantially influenced by the dominant users of data transmission services — namely, large multilocational corporations. Given national differences in industrial organisation and concentration, the strength of this monopsonistic influence varies between countries: in the UK, 60 per cent of data communications traffic is generated by 300 large companies; in the US, 50 per cent of telecommunications carrier revenues is generated by only 4 per cent of service users; in Norway, twenty-five companies account for 40–50 per cent of national data traffic.[3] As a result of these patterns of user concentration, the topological characteristics of private computer networks undoubtedly have a major shaping influence on the geography of infrastructure development and inter-city tariffs for data communications services.

Canadian evidence, at least, indicates that large multilocational firms as well as federal government agencies tend to implement highly centralised network topologies, with about 90 per cent of data processing capacity being concentrated in computer centres located at, or near to, head-office establishments (Hepworth, 1987a). The aggregate outcome of these private networking strategies, as they are conditioned by Toronto, Montreal and Ottawa's dominance of Canada's corporate and government head offices, is the skewed metropolitan structure of technological concentration shown in Table 8.3. In another paper, Hepworth (1986b), it was argued that the general effect of this concentration pattern is potentially to reinforce emerging 'central-periphery' relations in Canada's information economy owing to significant inter-urban tariff differentials in data communications services. The quantitative effects of these tariff differentials on Canadian metropolitan economies, as they impinge on comparative and competitive advantage, have not yet been assessed (or systematically investigated). However, given that telecommunications costs

Table 8.3 *Concentration of computer population in Canada by selected metropolitan areas, 1983*

Area	NATIONAL COMPUTER POPULATION		PROVINCIAL COMPUTER POPULATION	
	Share (%)	Concentration	Share (%)	Concentration
British Columbia: Vancouver	8.4	149	72.0	145
The Prairies:				
Calgary, Alberta	5.7	188	47.0	157
Edmonton, Alberta	4.9	156	40.4	130
Regina, Saskatchewan	1.5	189	49.6	239
Saskatoon, Saskatchewan	0.9	129	29.3	164
Winnipeg, Manitoba	3.5	143	90.9	152
Ontario:				
Ottawa-Hull	6.0	187	12.8	149
Toronto	25.0	174	53.1	139
Quebec:				
Montreal	11.9	100	70.3	140
Quebec City	1.8	84	10.9	125
Atlantic Provinces:				
Halifax, Nova Scotia	1.8	177	71.4	215
St John, New Brunswick	0.5	116	32.5	172
St Johns, Newfoundland	0.8	135	73.4	219

Note: Concentration Index calculated by dividing metropolitan share of provincial/national computer population by metropolitan share of provincial/national labour force.

Source: Hepworth (1987a).

make up (on average) less than 1 per cent of the total operating costs of large companies, it will be small- and medium-sized firms located in Canada's 'peripheral' regions which will be most adversely effected by differential urban accessibility in national telecommunications (Canada Department of Communications, 1987).

With the notable exception of Langdale's (1983) study of US telephone markets, there is an unfortunate dearth of research on the geography of telecommunications and its attendant implications for urban and regional development. As a result, we have failed to capitalise on the earlier work of Pred (1977) and Tornqvist (1973), Abler (1977), Pool (1977) and various other researchers, while continuing to speculate rather than investigate the real 'space-time convergence' effects of advanced telecommunications. In sum, we appear to have been standing still for at least fifteen years:

Electronic media will soon give us the power to create a communications world without distance, and in such a world, it will no longer be necessary together in small areas to communicate with like people. What a frictionless space, in conjunction with the shift to quaternary-oriented society, portends for general spatial organisation is unclear. [Abler, 1974: 341]

In the meantime, the research space left by geographers has been (inadequately) filled by scholars from other disciplines, judging from a recent multidisciplinary review of telecommunications literature (Snow, 1988).

The 'wired city': what's in a name?

Metropolitan cities are, of course, already 'wired up' with ordinary telephone lines. These communications media are already being used for various computer network applications, such as so-called 'tele-shopping' and 'telebanking', electronic funds transfer, inter-firm order entry, point-of-sale transactions, 'helpline' projects in the voluntary or non-profit sector and so on. Additionally, there are cable systems for television entertainment and information services, and even the electricity grid provides wires for telemetry applications; further, telex, the 'workhorse' of business communications, is still widely used for message exchanges. So, what is the 'wired city'?

The history of the wired city concept dates back to telecommunications policy debates in the US which arose in the context of the Johnson Administration's major urban programmes (Dutton, Blumler and Kraemer, 1987). At that time, the Goldmark panel was established by the National Academy of Engineering to advise on how advanced telecommunications could be used to promote local economic development and improve urban life more generally (Goldmark, 1972). Although the wired city concept was invoked in numerous cable projects across the US during the 1970s (Mitropoulos, 1983), its instrumental use by industry and governments has changed considerably over the course of this decade. As stated by Dutton *et al.* (1987: 7):

Early visions of wired cities embodied images of electronic democracy, education and huge lists of interactive services, with the public increasingly voting, shopping and even producing television from the home. Later versions continued to embody communications technology as an approach to electronic information and transactional services. However, the advanced wired city ventures evoke a more down-to-earth set of concerns over: (a) the provision of basic telephone, television and computer services with an already demonstrated value to households and businesses; (b) the growing significance of telecommunications to the fortunes of localities and nations; and (c) the furtherance of national interests and cultural values.

It is these 'later versions' of the wired city concept that are emerging as blueprints for urban investment and planning in the US, Europe and Japan. Wired city plans are still more apparent than real, and surrounding sales 'hype' (emanating from local govenments and business interests) tends to obscure their actual extent of implementation. Nevertheless, as grandiose plans for urban development in the 'information age', they are multiplying rapidly, and symbolic investments abound: Manchester has publicised its intention to become Europe's first 'on-line city'; Barcelona uses the label 'Telematics City'; Cologne has declared itself 'Communications City'; Amsterdam is 'Informatics City; and, Japan's plans for 'Advanced Information Cities' are well publicised.

The key driving forces of these comprehensive planning projects are threefold. First, city authorities perceive the creation of new technological infrastructure — teleports, fibre optic cable, home automation systems, 'intelligent' building complexes, etc. — as important for developing a comparative advantage at the national and international economic levels. Second, land and property developers are attempting to create markets in so called 'high-tech' real estate — 'smart' buildings, 'smart' parks, etc. (Sugarman, 1985; Vincent and Peacock, 1985). And, third, individual cities are being targeted as primary and 'showcase' markets for informatics products and services by national electronics industries and domestic telecommunications carriers. Reflecting this convergence of state and market interests, and notwithstanding their economic scale, technical complexity and political sensitivity, grandiose wired city plans are normally undertaken by investment consortia consisting of different levels of government, the electronics and telecommunications carriage industries, and various groupings with a stake in the urban land market (developers, planning and architectural practices, engineering consultancies, etc.).

The centrepiece of wired-city planning is increasingly the teleport. Teleport schemes, in general, include an international communications 'gateway', a metropolitan area network for distributing advanced information services and surrounding 'intelligent' building complexes. In the US, for example, forty-five teleports were operational, planned or under construction in North America by May 1986; in Germany, teleports in about eighty cities are planned to be interconnected with a nationwide fibre optic network by the mid-1990s (Noothoven van Goor and Lefcoe [eds], 1987).

The underlying rationale for teleport developments is remarkably similar across countries, although the dominant points of departure for local initiatives differ (Lange, 1987). In the Netherlands, for example, teleports in Amsterdam and Rotterdam are, respectively, planned to attract foreign multinational offices and to enhance the international competitiveness of the Dutch seaport industry. In the US, New York's teleport was initially marketed to the 'Wall Street' business community as a means of 'by-passing' the local telephone system and frequency congestion (Moss, 1986a); the City of San Antonio's links with the Texas teleport project were motivated by inter-city competition for 'high-tech' industry in the 'Sunbelt' region (Blazar, Spector and Grathwol, 1985). In France and Japan, teleport projects intersect with national 'technopolis' programmes, which are designed basically to achieve technological supremacy in the global economy (Glasmeier, 1988; Ihda, 1987).

Teleport projects, in most cases, are conceived as a new technological infrastructure for urban and industrial renewal. In Osaka and London, teleports are simply one infrastructural element (along with light rail transit and new roads, for example) of major urban redevelopment programmes being undertaken to 'revitalise' docklands areas. The major beneficiaries of these 'revitalisation' programmes are the real estate industry, as pointed out by Itoh (1987: 169):

> The Teleport concept has been just a well-timed support for the real estate society. The Teleport has allowed the real estate industry the opportunity to rethink the development strategy for the harbour area. In such sense, the Battery Park City project in New York seems to me an excellent example of a project connected to a Teleport concept.

The financial viability of teleport-centred land redevelopment depends on urban planning policies for redistributing economic activity, attracting new industry and a considerable marketing effort. In Tokyo and Osaka, for example, zoning restrictions are being used to deconcentrate offices from the central business district to 'intelligent' parks located in the teleport area (Huet, 1988); the New York project, similarly, includes plans for redistributing 'back office' employment from the high-rent business district of Manhattan to cheaper local labour and land markets in the outer metropolitan boroughs of Brooklyn, Staten Island and the Bronx (Moss, 1986a). The teleport projects in Amsterdam and Cologne are underpinned by specialised 'mini growth poles' within the central city, respectively called 'Informatics Zone' and 'Media Park'. In Toronto, by way of contrast, 'high-tech' office development in the old railwaylands area (south of the core business district) has been driven by market forces which run counter to the basic provisions of the city's structure plan for transportation land use (Hepworth and Dobilas, 1985).

The total effects on urban development of teleport projects are, in these embryonic stages, impossible to assess. What is clear, however, is that the first-phase 'winners' are large corporations — the high-volume users of international communications (particularly in financial services and media), telecommunications carriers, electronics manufacturers and dominant firms in the urban land and property market. What is not yet clear is how other sectors of the local economy — such as, small- and medium-sized firms and households — will be affected by teleport developments, particularly through the spatially differentiated impact (real and/or anticipated) of new forces operating in the urban land market. In the London Docklands case, at least, all the available evidence suggests that the business of 'landtronics' (capital combinations of land and the new information technology — 'smart' parks and buildings) is simply one aspect of a market-led urban renewal programme, which is fuelling local processes of social and economic polarisation (Nicholson, 1989).

Despite the financial and political risks surrounding teleport projects, as master plans for urban renewal, city authorities still appear to perceive these infrastructure developments as viable 'technopoles' in the inter-urban competition for economic activity (Abler, 1987). In this sense, teleports are seemingly the 'latest craze' in the recent scramble amongst city governments for 'high-technology' more generally (Gordon and Kimball, 1987). The risks associated with teleport investments are, however, higher and more complex because of their additional dependence on local demand for information and communication services by firms and household consumers.

The commercial development of business applications in local cable services, notably, does not require links to a teleport. In the UK, for example, cable companies in Coventry and Milton Keynes market on-line, business contact systems to local firms, with the basic aim of spatially containing linkages (Hepworth, 1988b). The systematic exploitation of information-based network externalities — using local telecommunications systems — is not far advanced in the UK, however, this may change dramatically as the wider diffusion of cable infrastructure converges with market-oriented developments in the municipal information economy (see below). In addition to reducing search costs, and reshaping 'information space' in favour of local suppliers, network externalities potentially accrue, for example from: order-entry applications, which lower inter-firm transaction costs; inter-organisational networking between related local industries (e.g. theatres, car hire companies and hotels), which lower marketing and distribution costs; and, shared access 'ports' to public data base services, which also act to lower search costs (Hepworth, 1988b).

Clearly, these network externalities available from shared-user computer networks (which can be constructed from ordinary telephone lines, and not necessarily broadband cable), represent new resources for local economic development policy — at least, in theory or technically. Indeed, this is the basic thrust of local cable network plans in the US, Europe and Japan. The social distribution of these externalities will, however, be determined much like other shared-user infrastructures (e.g. transport and energy systems), that is: by institutional arrangements related to ownership and control, and attendant pricing and investment policies. As Dutton, Blumler and Kraemer (1987) have emphasised, the 'wired city' could simply develop as a 'network marketplace' for information commodities, as a public utility replete with state-subsidised information, or as a 'mixed' economy of information transactions. As such, the 'wired city' would develop differently in different cultural, political, institutional and social settings.

The same conclusion applies equally to the so-called 'home information revolution', which Castells (1985: 16) sees as 'the most important consequence of new communication technologies for urban life and forms'. Several types of computer network applications are classifiable under this rubric, for example: telebanking and teleshopping; home security systems, wiring to local police, fire and ambulance stations; telemetry, for recording energy usage; interactive cable television, for capturing data on consumer responses to advertising; 'helpline' projects, as electronic 'clearing houses' for social services demand; community bulletin boards, teleschooling or 'distance learning', based on video disks; and so on. Several of these applications are already in common use, or are in market/technical trial phases of development (Miles, 1988b).

In essence, the 'home information revolution' is a new aspect of what Gershuny and Miles (1983) dub the 'self-service economy', characterised by the decentralisation of services production from firms (and governments) to households (see also Gershuny, 1978). Importantly, given that only a small share (less than 10 per cent) of household expenditure is devoted to consumer information services (e.g. television, newspapers and magazines), the creation and expansion of a mass home information market depends crucially on organisational and technological changes in production and distribution processes, for the economy taken as a whole. Thus, telebanking is really one aspect of production decentralisation in the banking industry; teleshopping is a form of electronic quasi-integration between retailers and consumers, rather than merchandise suppliers; 'tele-' activities in public services are similarly new, flexible, decentralised, structures of services production and distribution used by governments. That is to say, the 'home information revolution' is integrally related to other 'information revolutions' in the economy, whether in the factory, the shop, the office or elsewhere (Hilbig and Morse, 1988; Robins and Hepworth, 1988a).

At this stage, the 'home of information revolution' appears to be accessible to only a small minority of so called 'Technologically Advanced Families'[4] judged even by Mitsubishi's market propaganda. The target market for the Japanese company's 'Home Automation System' potentially comprises 17 per cent of US households, who part-time 'telework' and own the following staple technologies; three television sets, one video-cassette recorder, one personal computer and a telephone. As such, for the large majority of urban households, the 'smart house' (dubbed '*domotique*' by the French telematics industry) is no more than a 'push-button fantasy' (Mosco, 1982), but whose importance as a mass market product innovation to electronics manufacturers and telecommunications carriers is becoming transparently obvious (Van Rijn and Williams [eds], 1988).

The municipal information economy

Municipal governments are emerging as major actors in the information economy. Their present involvement in the creation of 'wired cities', through telecommunications, teleports and 'high-tech' zoning, for example, is basically a continuation of the historical role played by local authorities in infrastructure planning and development.

In a recent survey of UK local authority policy initiatives in the area of economic development, Sellgren (1987) found that the provision of information to firms and individuals was a rapid growth area of municipal services. The survey evidence indicated, for example, that 'in 1983 only 38.5 per cent of district level local authorities in Great Britain were involved with offering business advice, yet by 1986 this had grown to 75.8 per cent' (p. 66). Parallel to this growth of information provision is the recent increase in local authority investments in data processing and communication. According to Grimshaw and Haddad (1988), the total annual budget of (400 plus) local authorities dedicated to information technology had reached £0.5 billion in 1988, and the municipal computer population was growing at an annual rate of 100 per cent, compared with 82 per cent for UK industry as a whole. These summary data suggest, therefore, that the importance of information resources is increasing in UK local government, but why?

The increased provision of information services to firms has come about for a number of reasons. First, local authorities increasingly view information as a saleable commodity, and as a new source of revenue at a time of fiscal austerity. Electoral registers are, for example, sold by most authorities to public data base vendors, whose 'on line' information services are then resold to market research companies. The growth of these market transactions are, then, part of the general expansion of the 'geo-information system' business (Openshaw and Goddard, 1987; Information Technology Advisory Panel, 1983).

Second, under the 1988 Local Government (Reform) Act, authorities are now required to 'create markets' for a growing range of municipal services (e.g. street cleaning and school catering), which they themselves previously provided. Central to this contracting out process — or the organisation of market transactions — is the provision of variegated information to potential suppliers — for example, information on product standards, tendering procedures and other contract-related information (Laffont, 1987). The Thatcher administration plans to increase the scale of contracting out in municipal services, by extending the range of services subject to the Local Government Act's relevant provisions, and by intensifying financial pressure on local governments through the new Community Charge (Noel Hepworth, 1988).

Third, and relatedly, entrepreneurialism is replacing managerialism as the leitmotif of local authority development policy (Harvey, 1987). A major result of this ideological shift, under the Thatcher administration, is increased economic competition between 'entrepreneurial cities'. As Sellgren's (1987) survey of local economic initiatives showed, municipal expenditures on advertising are growing significantly as authorities compete for industry and employment at the international and national scales. The rationale of these policy directions was highlighted ten years ago by Pred (1977), particularly in the context of corporate 'information space' and its potential reshaping through regional policy initiatives:

> This alternative would require that government and planning authorities establish an agency to subvert the limited-search syndrome by actively providing decision-making units with detailed opportunity — and cost-specific information on explicit and implicit locational options. The information supplied would be much more refined than that normally found in glossly promotional booklets and would in effect represent a public subsidisation of search costs. The information would also be designed to reduce uncertainties about the viability of operating in 'backward', 'lagging' or 'depressed' regions. If effective with respect to implicit location decisions, the information provided would help divert many employment multiplier effects away from large metropolitan complexes' [p. 207].

In the UK, at least, information-based economic policy in local authorities covers a wide range of activities, such as: sales and marketing support to small business; participation in the burgeoning 'conference business'; trade missions to the US, Europe and the Far East; official delegations to the European Economic Commission

(EEC) and local EEC offices for investment funds; 'sales' offices in London to attract local firms 'northwards'; subsidies to local audio-visual industries and corporate video, in particular, as a basis for enhancing the 'image' of so called 'backward' regions; and so on.

Parallel to the rise of these quasi-markets for local authority information services (i.e. producer services as 'public goods') is the increased provision of information to individual consumers and service users. The recent growth of these informational activities is fuelled by similar political and economic forces that affect all local authorities in the UK. First, advertising and promotional activity is central to the promotion of local consumer services — theatre, restaurants, 'heritage' tourism, retail, etc. — and to the financial viability of large-scale shopping malls and entertainment centres (Harvey, 1987). Second, the privatisation of municipal services — through asset sales, 'opting out' arrangements and contracting out, generally — has forced local authorities to compete with private firms in markets for education, manpower training, housing, etc. In this incipient context of 'consumerism' in municipal services, the provision of information to customers/clients is viewed as critical to competitiveness (Hambleton, 1988).

These informational activities have, however, intensified for more obvious political reasons. With what Moore and Booth (1986) call the 'privatisation of urban policy' — that is, the by-passing effect of private sector–central government initiatives, such as Urban Development Corporations and Enterprise Zones — local authorities are using information services to increase public support for, or to legitimate, their historical role in urban governance (Hambleton, 1988). At the local level, public criticism of municipal government operations — for their 'ivory tower' remoteness and bureaucratic inflexibility — has reinforced central government demands for higher levels of efficiency in services delivery and management (Whynes, 1987).

In addition to investing in self-promotion campaigns, local author-ities have responded to these dual pressures for greater fiscal and political accountability by 'decentralising' their basic operations. At present, with the Community Charge imminent, these organisational reforms have accelerated throughout the UK, and all are intended to push managerial responsibility (and financial accountability) for local services provision 'down the hierarchy' and 'away from the Town Hall', and in the general direction of the 'market'. These decentralised units may, for example, be multiservice neighbourhood offices (e.g. in Islington and Birmingham), 'generic' area teams (e.g. in East Sussex) or 'patch' teams for residential and home-based social services. (See, for case histories, Morphet, 1987; Elcock, 1986). The following example relates to the Borough of Walsall's 'decentralisation' plans for housing services, through the creation of thirty-one neighbourhood offices in 1981. As described by Baddeley and Dawes (1987), the rationale for these organisational reforms was highly complex in terms of political ideology:

> The Walsall politicians in 1981 saw the way services were being delivered as potent sources of public passivity or its equally ineffectual corollary, violence. Their vision was to transform the dependency induced by exclusion from the market into a mature involvement in the political process. Their means involved dramatic organisational change to get closer to the authority's citizens through physical location, scale and layout of buildings. Once this was achieved they wanted to present information as straightforwardly as possible to people through staff, through posters and newsletters and through IT (Information Technology). The idea that 'information is power' was real for them and they wanted the organisation and its technology to help staff to give information in ways which would reduce dependency and empower people. [p. 2]

The rising level of information technology investments is, then, integrally related to broader processes of organisational change — 'decentralisation' both within local authority organisations and to markets — which affect all municipal governments in the UK. These technological innovations are, in theory, expected to resolve the recurrent tension between 'centralisation' and 'decentralisation' in organising and managing local government. For example, in Walsall's case, the authority's technological strategy is designed to meet the following (technical) objectives:

> The technology that is wanted is one that maximises the capacity of neighbourhood offices to exercise discretion on the basis of local experience without the delay that attends the search for authorisation from the centre. The technology should provide on-screen access to the routine details of the progress of requests from customers and the mechanisation of routines procedures. At the same time information generated in the neighbourhood should be capable of being aggregated and monitored at the centre [Baddley and Dawes, 1987: 12].

The UK situation is not presented here as having general applicability to local governments in other countries. There are, however, certain parallels with developments in the US, where the large-scale contracting out of municipal services occurred much earlier (Moore, 1987) and the creation of 'little city halls' arose from decentralisation political pressures which intensified during the 1960s (Starr, 1987). In the US context, Kraemer and King (1988) also highlight the flexible capabilities of information technologies for supporting decentralised and centralised structures of organisation and decision-making. Their view of the changing role of information technology in city governments is summarised in Table 8.4.

Table 8.4 *The role of information technology in government*

	1950–1980	1980–2000
Technology	Computers	Computers, telecommunications, and management science techniques
Application	Mainly internal to the government and focused on operational performance of the bureaucracy	Strategic use of information technology; external focus on relation of government to its clients; internal focus on planning and control by city management
Impacts	Productivity in large volume information handling; quicker, more informed decisions about internal operations; stable government size despite greater service demands; bureaucratic politics concerning control of computers and internal use of information	Productivity in shared use of information and direct citizen access to service systems; quicker, more informed decisions about external services; stable government size despite greater citizen access and increased 'tailoring' of services; bureaucratic politics concerning control of systems and external use of information

Source: Kraemer and King (1988), p. 34.

Of particular interest is Kraemer and King's view of the 'new opportunities' for revenue expansion presented to local governments which derive from modelling municipal computer network applications on the same line as airline reservation systems. (See the Air Canada example in Chapter Five, and the same chapter's discussion of electronic 'quasi-integration'.) In the case of local authorities, it is suggested by Kraemer and King (1988: 39) that

> City governments could develop similar relationships with citizen and corporate clients through strategic use of their computerised systems to:
> (1) share information collected from citizens back to them; and
> (2) facilitate direct access by citizen and corporate clients to government information collections.
> For example, cities could directly share the public data in their real property information systems with title insurance companies, real estate agents, property appraisers, and developers-builders thereby reducing the costs of doing business for these firms and raising revenues for the government by collecting fees for access to the data. [p. 39]

Although local authorities in the UK, at least, are not yet making a significant business out of information, it is extremely plausible that they will move in that general direction. This may come about from the convergence of four basic forces, namely: the search for new sources of revenue; the significant shift of the local authority functions from services production to management; the growing tendency towards 'entrepreneurialism' and 'professionalisation' in municipal information services; and the technical capabilities of computer network systems for codifying and costing the information component of municipal service transactions (Graham, 1988).

The general result of the operation of these political, economic and technological forces would be, as Aglietta (1979) suggests, to reinforce present tendencies towards the privatisation of municipal services according to 'information principles'. On the one hand, the codification of transaction-related information enables local authorities to control effectively for the 'public-good' characteristics of municipal services — that is, the information-based externalities associated with market failure (so called 'demand-revelation problems') can be internalised; on the other hand, information captured through market transactions, or the process of consumption, can be itemised and billed, as well as resold to the local authority's corporate clients as market research data. These technical processes and their central role in the information economy form the basis of what Mosco (1988) calls the 'Pay-Per Revolution'.

The creation of a 'network marketplace' (Dordick, Nanus and Bradley, 1981) for local authority information would create a new basis for class divisions within cities: the 'information rich' (those able to pay for information); and the 'information poor' (Demac, 1988). This possibility is underlined by Golding and Murdock (1986), based on their reporting of studies carried out by the UK's National Consumer Council, as follows:

> There is evidence to suggest a massive unmet need for information and advice provision. Such changes have a disproportionately large impact on lower income groups, who are more dependent on local government for information, advice and leisure facilities. [p. 84]

These emerging aspects of social inequality in the so-called 'wired city' are, of course, reinforced by other basic conditions which affect access to the information economy, such as: the price of 'home automation' technology (Bruce, 1988); but, perhaps more fundamentally, the unequal distribution of variegated knowledge and educational opportunities in society more generally (Gandy, 1988). In addition to this consumption-based form of closure, the municipal information economy is effectively being 'ring-fenced' through the process of privatisation. According to Starr (1987: 132):

> Privatisation does not transform constraint into choice; it transfers decisions from one realm of choice — and constraint — to another. These two realms differ in their basic rules for disclosure of information: the public realm requires greater access; private firms have fewer obligations to conduct open proceedings or to make known the reasons for their decisions. The two realms differ in their recognition of individual desires; the public realm mandates equal voting rights, while the market responds to purchasing power. They differ in the processes of preference formation: democratic politics is a process for articulating, criticising, and adapting preferences in a context where individuals need to make a case for interests larger than their own. Privatisation diminishes the sphere of public information, deliberation, and accountability — elements of democracy whose value is not reducible to efficiency.

Thus, when we situate the emergence of the 'information city' in the present context of political, fiscal and organisational change in UK local government, the utopian scenario of an 'electronic democracy' lacks any credibility (see, for example, Toffler, 1981: Martin, 1978; Lowi, 1980). It seems implausible that the same 'airline reservation system' being used to propel local authorities into the information business would also serve as a communications channel for mobilising public interest groups. While the new technology and the 'information explosion' may, indeed, lower the transaction costs of mobilising private and public interest groups in the political process, as Kling (1988) argues, we should bear in mind that the municipal information economy is presently being reorganised around market forces.

Conclusion

Metropolitan cities are the 'backbone' of the information economy at the national and global levels. As 'wired cities', they are 'targets' for the information technology industries as much as being arenas for new types of urban regeneration initiatives. They are also the major 'crossroads' of the new 'electronic highways', whose planning and development are critical to inter-urban competition for wealth and employment in the information economy. Importantly, in countries where 'deregulated' telecommunications policy regimes operate, the construction and pricing of 'electronic highways' are fundamentally determined by computer network innovations in large corporations and the profitability imperatives of private 'natural monopoly' carriers — such that urban or regional policy considerations are insignificant or secondary. In some countries, such as France and Japan, telecommunications policy has a strong urban component, due largely to the central role that cities in these economies are planned to play in national 'high-tech' industrial strategies.

Clearly, the 'information city' is also developing as a place where social polarisation tendencies are likely to be most extreme. We noted that metropolitan cities, such as Toronto and London, are highly specialised in information production and exchange, but Chapter Two also revealed that 'outside' the information sector new job opportunities appear to be restricted to lower order service occupations. These class inequalities in *production* are, importantly, converging with emerging inequalities in *consumption*, as the operation of what Aglietta (1979) calls 'information principles' threatens to undermine the local welfare state and possibly local democracy. The current dearth of urban research on the information economy, however, is manifestly obvious. In some cases, geographers have lapsed into hyperbolism, as Castells's (1985) following dystopian vision of the 'informational city' clearly indicates:

> There will remain switched-off, wireless communities, still real people in real places, yet transformed into urban shadows doomed to haunt the ultimate urban dream of the new technocracy. [p. 19]

Importantly, even this Cassandra-like warning issued by Castells is rendered insignificant by the sales hyperbole of not only dominant sections of the information industries but also by the promotional activities of entrepreneurially minded local governments.

In writing this book and publicising my ideas on urbanisation processes in the information economy, I am highly conscious of the generality of much — if not all — of the discussion. For most of my peers in the discipline, 'real geography' is the stuff of places (Massey, 1984) and their unique experience of capitalist development through time. Reflecting this orthodoxy, my current research on 'urban development in the information economy' is concerned with the interplay of local development forces with the 'megatrends' described in this book. What does the 'information city' look like in London, Bristol, Manchester, Newcastle, Leeds, Birmingham and other places?

Notes

1. California Legislature, Assembly Bill No. 3487, 17 February 1988.
2. Japan Ministry of Posts and Telecommunications, 'The Present Structure of the Japanese Telecommunications Sector', Paper presented at the *Third Meeting of Japan–UK Telecommunications Regional Consultations*, London, November 1985.
3. See the *Economist, Survey of Telecommunications*, 17 October 1987.
4. 'Home Automation Update', Mitsubishi Electric Sales America, Vol. 1, No. 1, April 1987, published by Medama.

Chapter Nine
Geography: Information at a Price

These are the golden years of the 'data merchant', the information broker who peddles his 'on-line' wares in the 'network market-place' (Roszac, 1986; Dordick, Nanus and Bradley, 1981). Witness the junk mail of the 'corporate census'! In exchange for token prizes or free entry into prize draws, we might reveal information about our income, occupation and consumption habits to mail-order houses, credit companies or manufacturers, and all of these details are coded to the lowest possible level of geographical disaggregation: our homes. From these 'maps', and others collected through everyday transactions (e.g. buying a car or taking out a bank loan), firms in different sectors are able to target consumers and plan their productive, distributive and marketing operations with greater precision. That is to say, information space has become a commodity in its own right, whether sold by private firms or by government agencies: its particularisation and commoditisation represents the spatial dimension of what Mosco (1988) calls the 'Pay-Per Revolution'.

This latest stage of codifying information space for commercial profit, at the fine detail made possible by computer network innovations, is the most marketable of the applied geographies spawned by the development of the information economy. As geographers, we are the 'data merchants' who specialise in the production and resale of made-to-order space, and our principal customers are firms and governments who can afford to pay for the information commodities we purvey. Our latest 'de-luxe' model is, of course, geographical information systems.

What lies behind these spatially intensive forms of 'data farming'? Why industrialise information space? In seeking answers to these questions, we must bear in mind that there is nothing new about these commercial developments, as Beniger's (1986) list of 'control innovations' clearly shows (see Figure 1.1). What does, however, make today's 'information business' significant historically is that it represents, along with the rest of high-technology industry, an important new area of capital accumulation in advanced economies (Castells, 1985). According to Schiller (1984), for example:

> The multiplying informational activities and the growing stock of instrumentation around us are attributable, in large part, to the economic, political and cultural crisis of the world market system. Information and information technologies have been seized upon as the means to alleviate and overcome the crisis. [p. xii]

In the course of this book, we have encountered various dimensions of this 'crisis' in different areas and at different levels of the capitalist economy, through our explorations of the geography of information and information technology. For example, in the firm-level case studies presented in Chapter Five, it was shown that computer network innovations offer large, multilocational corporations a 'way out' of declining 'home' markets (for example, Canadian regional markets in the case of the *Globe and Mail* newspaper, national markets in the case of Bell-Northern Research). Similarly, the globalisation of capital markets permitted by information technology has been crucial to the profitability requirements of large financial conglomerates (Chapter Seven), and probably also to productive decentralisation in multinational corporations during the recent period of instability in the world monetary system. Again, in the UK currently, information and information technologies are perceived by local government as vital to their very survival, in the face of the Thatcher administration's financial and ideological onslaught (Graham, 1988). At the same time, new planning concepts, such as the 'wired city' (originating in the US), and the 'advanced information city' (from Japan), are being grasped by local governments as a new formula for inner-city regeneration, but also by powerful information industries and property developers whose profitability imperatives lie behind the 'wiring up' of cities all the way to our living-rooms (Hepworth, 1988b).

In Chapter Six, I provided an information economy perspective on the geography of flexible production, which is thought to arise from the current crisis of 'Fordist' economic and political institutions. In this discussion, I suggested that the new-found 'flexibility' of manufacturing firms is traceable to the growing information intensiveness of their total operations, as reflected in the increasing centrality and closer articulation of production engineering, R and D and marketing functions, and, more broadly, the changing composition of production factors, in favour of 'information capital' and 'information labour'. With regard to 'just-in-time' innovations, my interpretation is that they represent a further aspect of this factor substitution process, whereby information replaces material capital (here, inventory) in inter-firm production linkages.

The basic implication of these factor-substitution processes, I suggested, is that the profitability of firms depends increasingly on being able to achieve *scale* economies in information rather than physical capital. Firms can not 'live' on economies of *scope* 'alone': we are not witnessing the 'end' of economies of scale, as a basic principle of productive and industrial organisation, but a fundamental change in its character, as firms grow more information intensive rather than materials and energy intensive. (Perez, 1985) The key technologies permitting informational economies of scale are, of course, computer networks which integrate functions across the entire firm — from factory production, to head-office management, to R and D laboratories, to distributive and marketing operations.

In this same chapter, I went beyond the 'flexible firm' to regional systems of flexible production, which some geographers see as manifestations of a 'post-Fordist' space economy in the making (Scott, 1988). I focused attention on the role of transaction costs in vertical (dis)integration, given that geographers have invoked this increasingly influential body of economic theory to explain 'flexible specialisation' in spatial context. Here, by highlighting the fact that the resources firms set against transaction costs are basically informational, I show why information linkages are thought to provide the 'glue' in spatial agglomerations of 'flexible specialisation'.

However, it was emphasised that transaction-cost economics does not go even halfway to explaining why firms (dis)integrate, and, if its analytical power is to be exploited to the full, then we must go beyond the firm or the sector — to *all institutions* involved in the complex process of 'making markets' in the regional economy. These institutions include industry associations, labour unions, consumer organisations and the local and central state — all of whose regulatory activities have a direct bearing on the level of transaction costs and, more importantly, how these costs are distributed between firms (large and small), between firms and government, between capital and labour, between labour and government, between different parts of the labour force, and also between producers and consumers. Consider, for example, Brusco's (1982) discussion of 'the Emilian model' of flexible specialisation and the central role of various institutions in securing informational economies of scale *at the level of the region*:

> Even if it is accepted that for many industries the importance of technical economies of scale has been overstated in the past, it might still be objected that there exist nonetheless both indivisibilities in the administrative work of firms and significant pecuniary economies of scale. Thus small firms might experience difficulties in book keeping, in obtaining raw materials, and in obtaining credit at the same price paid by larger firms with greater bargaining power. But in this context, it is extraordinary to observe how the artisans and small entrepreneurs of Emilia-Romagna have overcome these difficulties by creating associations to provide these purchasing and credit negotiations, thus establishing on a co-operative basis the conditions for achieving minimum economic scales of operation. These associations, which cover the whole region, prepare the pay slips, keep the books, and pay the taxes of the small firms, giving to the latter the expertise of a large office in administration and accountancy at a minimal price. Furthermore, these associations also establish technical consultancy offices, consortia for marketing and the purchase of raw and semi-fabricated materials and, most importantly, cooperatives which provide guarantees for bank loans which can thus be obtained at the lowest possible rate of interest. [p. 173]

In other words, to do justice to the 'transaction-cost' framework, it is necessary to specify the entire institutional fabric of local economies, before adumbrating a new theory of regional development based on 'post-Fordist' modes of production (Clark, Gertler and Whiteman, 1986). We saw, for example, that the growing informational activities of local authorities in the UK directly impinge on the transactional structure of the economy; further, 'vertical disintegration' (contracting out) in these authorities has very little to do with the fuzzy calculus of transaction costs, and more to do with the political 'New Right' ideology of the present Thatcher administration. Similarly, where these types of costs are incurred by Japanese manufacturing firms subcontracting to local component suppliers in the UK, they derive ultimately from the 'local content' provisions of EEC legislation pertaining to foreign direct investment which have broad political, economic, social and cultural objectives.

Indeed, much of the information economy consists of 'making markets' and, as Chapter Eight on financial services illustrated, information technology is playing a key role in this process. This includes not only in-house corporate activities — such as advertising, sales and purchasing, and market research itself — but also an entire array of new 'on-line' information industries (see the I.P. Sharp case study in Chapter Five), as well as specialised associations. Additional to these private sector institutions are, of course, the various levels of government, from the central regulators of market competition to the local regulators of the urban land market.

It is obvious that the re-ordering of markets — including their geography — cannot go forward without information. In the UK, at least, the most glaring evidence of this is the boost to the information industries forthcoming from the consolidation of the EEC market (the 'run-up to 1992'). These major institutional developments have generated a 'bonanaza' for university-based experts, the media and law and accountancy firms, and also revitalised government bureaucracies not just in Brussels but in the different national civil services where the minute details of 'market harmonisation' have to be carefully worked out.

In the UK, and in other European countries, the information economy has received a further boost from 'deregulation'. As Loasby (1976) notes, 'competition implies ignorance, as non-economists have always believed: competition is not a state of equilibrium but a process of search' (p. 184). Importantly, for our consideration of geography, 'old' information is public, spatially concentrated and can be readily found in the publications, files and archives of public bureaucracies; but 'new' information is private, dispersed throughout the space economy and it needs to be retrieved from individual firms and consumers, if new markets are to be established at all. It is this search for new information, fuelled by market deregulation and internationalisation, which accounts for the burgeoning development of 'on-line industries' worldwide: the marketability of information space and its delivery through 'geographical information systems' has increased accordingly.

Equally, and perhaps less profoundly, to do justice to the role of information technology in shaping the geography of economies, it is necessary to consider the combined power and effects of computers and telecommunications. In Chapters Three and Four, I explained how these innovations can be viewed as 'companion' technologies — by making computer networks the technical focus — and why this conceptualisation matters for modelling the spatial dynamics of economies. Here, I have suggested what theoretical considerations might be brought forward in analysing the role of telecommunications in regional development, by highlighting the 'communicability of information capital'. The complex questions raised by invoking this concept comprised only a critique of conventional regional models; there remains the task of theoretical reconstruction. Nevertheless, it is clear that, where information technology is concerned, we cannot logically have separate geographies of 'communications' and 'flexible production', nor should our consideration of telecommunications be concerned exclusively with office location and the 'future' of metropolitan areas. Computer networks, as a new form of information capital, reach the heart of manufacturing production these days.

A major implication of the communicability of information capital is that it portends a new phase of industrial relations, in which the bargaining power of employers is considerably strengthened (Aglietta, 1979). Not only do the new computer network technologies provide for higher degrees of decentralisation in the labour process (with so-called 'teleworking' being the most extreme mode of work organisation) but also they enable management to control 'remotely' and co-ordinate centrally geographically dispersed components of the 'electronic workplace' (Robins and Hepworth, 1988b). In this respect, the information economy is the economy of the 'flexible worker', whose collective bargaining strength is imperilled by a quantum change in the differential mobility of capital and labour.

We encountered some evidence of the 'reach' of computer networks in Chapter Five, through some firm-level case material drawn from my doctoral research, and observed the dramatic effects these innovations may have potentially on the inter-regional geography of production, markets and work. As I emphasised, these particular case studies had the limited objective of empirical description rather than of revealing the traumas of a 'new' industrial geography in the making. Nevertheless, if we lay this rather modest case-study evidence alongside real developments in the world's capital markets, our critique of regional capital theory, the claims of post-Fordist industrial geographers, the powerful market thrust of the 'IT' industries, then it is difficult to avoid coming to the following conclusion: a new geography of the economy is unfolding, and to understand its structure and dynamics, information and information technology must be brought to account.

In order to develop an understanding of the 'new' geography of enterprise, it will be necessary to 'get inside' the firm and to 'stay inside' in order to carry out the type of longitudinal analysis which the innovation process demands. Once 'inside', consideration should be given both to the geographical and *social* location of information, and to how the division of labour within and between firms is determined by the'locational' dynamics of the new technologies. For this purpose, geographers might profit from work in other disciplines — for example, organisation theory, the sociology of knowledge, psychology, the economics of information, communications sciences and labour process studies. The multifaceted nature of information, both as an economic and social relation, calls for a multidisciplinary approach to the changing geography of enterprise being brought about by the impact of information technology on organisations. What are the *limits* to 'flexibility'? To answer this question adequately, we need to see the firm not as 'production function' or the market as an 'environment', but both of these central objects of economic geographical analysis as institutions (Melody, 1987). If we do make this change in our vision, their informational dimensions become clearer, and with it the role of information technology in the spatial restructuring of economies is seen less 'through a glass darkly'.

In my view, the future value of geographical research to public policy debates hinges on whether we can *rapidly* develop a sound understanding of the spatial aspects of the information economy. This concluding statement is surely justified by the *current* division of labour in industrialised countries, whereby the ascendancy of the so-called 'information worker' appears to be going hand-in-hand with the 'disappearance' of the 'old factory worker', with the rest of the labour market being propped up by an abundance of 'flexible' job opportunities, a dismal array of lower order service occupations and, in the UK at least, by the Thatcher government's numerous training schemes for the 'employable' and 'unemployed'. What makes these labour market trends even more disturbing is the information economy's penetration of the welfare state, where the 'information principle' operates to create a new basis for class inequalities in the local community.

Appendix A
Inventory of Information Occupations

The table below provides the detailed listing of information occupations used by the Organization for Economic Co-operation and Development (OECD, 1981, 1986). It is based on Porat's (1977) original taxonomy, but adapted to facilitate international comparisons. The occupational codes are from the International Standard Classification of Occupations (revised edition 1968, International Labour Organisation, Geneva); and there are 284 'unit groups' covering 1 506 occupational categories.

I. INFORMATION PRODUCERS

Scientific and Technical

0-11 Chemists
0-12 Physicists NEC
0-13 Physical Scientists NEC
0-22 Civil Engineers
0-23 Electrical and Electronic Engineers
0-24 Mechanical Engineers
0-26 Metallurgists
0-27 Mining Engineers
0-28 Industrial Engineers
    (Except 0-28.30)

0-29 Engineers NEC
0-51 Biologists, Zoologists and related
0-52 Bacteriologists, pharmacologists
0-53 Agronomists and related
0-81 Statisticians
0-82 Mathematicians and actuaries
0-90 Economists
1-92 Sociologists, anthropologists and
    related

Market Search and Co-ordination Specialists

4-10.20 Commodity Broker
4-22   Purchasing agents and buyers
4-31   Technical Salesmen and advisors

4-41   Insurance and stock agents,
        brokers and jobbers
4-42   Business Services/Advertising
        Salesmen
4-43.20 Auctioneers

Information Gatherers

0.28.30 Work Study Officers
0-31   Surveyors (land, mine,
       hydrographic, etc.)

0.33.20 Quantity Surveyors
4-43.30 Valuation Surveyors

1-39.50	7-54.70	
3-59.30	8-59.20	Inspectors, viewers and testers (various)
3-59.45	9-49.80	
3-91.50	5-89.20	Information Gatherers NEC

Consultative Services

0–21	Architects and town planners
0–32	Draughtsmen
0–61	Medical practitioners
0–69	Dietitians and Nutritionists
0–75.20	Optometrist
0–83	Systems Analyst

0–84.20	Computer programmer
1–10	Accountants (except 1–10.20)
1–21 and	
1–29	Barristers, Advocates, Solicitors, etc.
1–39.20	Education methods adviser
1–62	Commercial artists and designers

Information Producers NEC

1–51.20 Authors

1–71.20 Composers

II. *INFORMATION PROCESSORS*

Administrative and Managerial

1–22	Judges
1–39.40	Head Teachers
2–01	Legislative Officials
2–02	Government administrators
2–11	General Managers

2–12	Production Managers
2–19	Managers NEC
3–10	Government Executive Officials
4–00	Managers (Wholesale/retail trade)

Process Control and Supervisory

0–33.40	Clerk of works
0–41.40 and 0–42.30	Flight and ship navigating officers

3–52	Transport and Communication Supervisors (except 3–59.30 and 3–59.45)
3–91.20	Dispatching/Receiving Clerk

3–0   5–20   6–00.30   Supervisors: clerical, sales and
4–21   5–31.20 6–32.20   other
7–0   Supervisors and General Foremen (production)

Clerical and Related

1–10.20	Auditor
3–21	Stenographer, typists and teletypists (except 3–21.50)
3–31.10	Bookkeeper (general)
3–31.20	Bookkeeper (clerk)
3–39.20	Cost computing clerk
3–39.30	Wages Clerk
3–39.40	Finance Clerk
3–91.30	Stock records clerk
3–92.20 and 3–92.30	Material and production planning clerks

3–93	Correspondence and reporting clerks
3–94	Receptionist and Travel Agency clerks
3–95	Library and Filing Clerks
3–99.20	Statistical Clerk
3–99.30	Coding Clerk
3–99.40	Proof reader

III. *INFORMATION DISTRIBUTORS*

Educators

1–31	University and Higher Education teachers
1–32	Secondary teachers
1–33	Primary teacher

1–34	Pre-primary teachers
1–35	Special education teachers

Communication Workers

1–51.30 Journalists and related
   and   writers (except 1–59.55)
1–59
1–73.30 Stage Director
1–73.40 Motion picture, radio,
        television director

1–73.50 Storyteller
1–74   Producers, performing arts
1–79.20 Radio, television announcers.

IV. *INFORMATION INFRASTRUCTURE OCCUPATIONS*

Information machine workers

1–63   Photographers and Cameramen
3–21.50 Teleprinter operator
3–22   Card and tape-punching machine
        operators
3–41   Bookkeeping and calculating
        machine operators
3–42   Automatic Data-processing
        machine operators
3–99.50 Office machine operators
8–49.65 Office machine repairmen
8–62   Sound and vision equipment
        operators

9–21 Compositors and type-setters
9–22 Printing Pressmen (except 9–22.70)
9–23 Stereotypers and Electrotypers
9–24 Printing engravers
     (except 9–24.15 and 9–24.30)
9–25 Photo-engravers
9–26 Bookbinders and related
9–27 Photographic processors

Postal and Telecommunications

3–70 Postmen, mailsorters, messengers
3–80 Telephone operators
8–54 Radio and television repairmen

8–56   Telephone and telegraph installers/
        repairmen
8–57.40 Telephone and telegraph linesmen
8–61   Broadcasting station operators

Appendix B
Network Diagrams

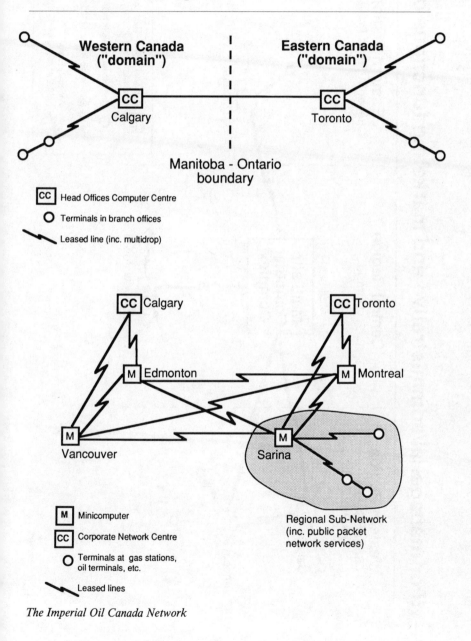

Western Canada
("domain")

Eastern Canada
("domain")

CC
Calgary

CC
Toronto

Manitoba - Ontario
boundary

CC	Head Offices Computer Centre
O	Terminals in branch offices
⚡	Leased line (inc. multidrop)

CC Calgary

CC Toronto

M Edmonton

M Montreal

M
Vancouver

M
Sarina

M	Minicomputer
CC	Corporate Network Centre
O	Terminals at gas stations, oil terminals, etc.
⚡	Leased lines

Regional Sub-Network
(inc. public packet
network services)

The Imperial Oil Canada Network

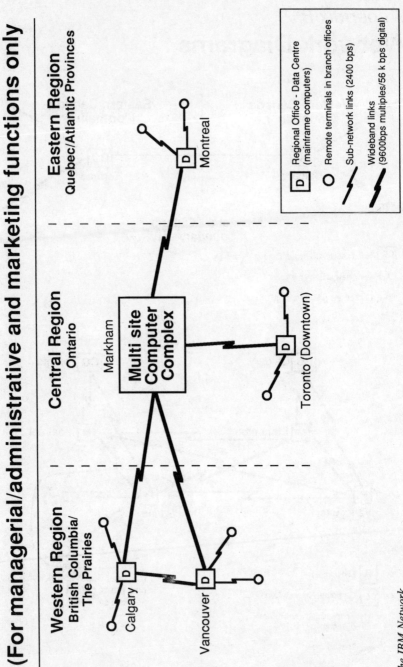

(For managerial/administrative and marketing functions only)

Western Region
British Columbia/
The Prairies

Central Region
Ontario

Eastern Region
Quebec/Atlantic Provinces

Markham

Multi site
Computer
Complex

Calgary

Vancouver

Toronto (Downtown)

Montreal

D Regional Office - Data Centre
   (mainframe computers)

O Remote terminals in branch offices

   Sub-network links (2400 bps)

   Wideband links
   (9600bps multiples/56 k bps digital)

The IBM Network

Operating regions

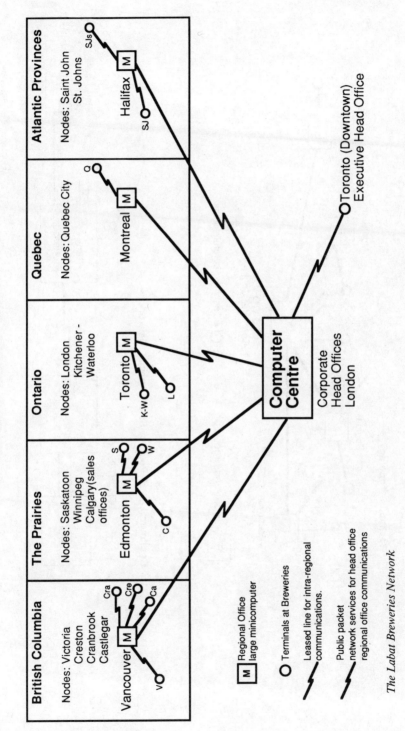

British Columbia	The Prairies	Ontario	Quebec	Atlantic Provinces
Nodes: Victoria Creston Cranbrook Castlegar	Nodes: Saskatoon Winnipeg Calgary(sales offices)	Nodes: London Kitchener - Waterloo	Nodes: Quebec City	Nodes: Saint John St. Johns

Vancouver **M** Cra Cre Ca V

Edmonton **M** S W C

Toronto **M** K-W L

Montreal **M** Q

Halifax **M** SJs SJ

Computer Centre

Corporate Head Offices London

Toronto (Downtown) Executive Head Office

M Regional Office large minicomputer

O Terminals at Breweries

⚡ Leased line for intra-regional communications.

⚡ Public packet network services for head office regional office communications

The Labat Breweries Network

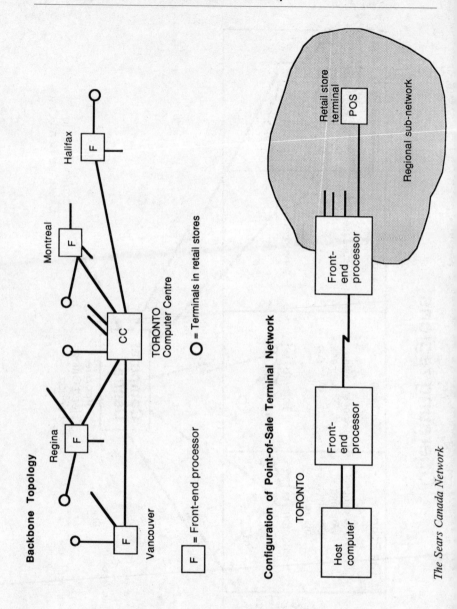

Backbone Topology

Halifax

F

Montreal

F

TORONTO
Computer Centre

CC

Regina

F

Vancouver

F

F = Front-end processor

O = Terminals in retail stores

Configuration of Point-of-Sale Terminal Network

TORONTO

Host
computer

Front-
end
processor

Front-
end
processor

Retail store
terminal

POS

Regional sub-network

The Sears Canada Network

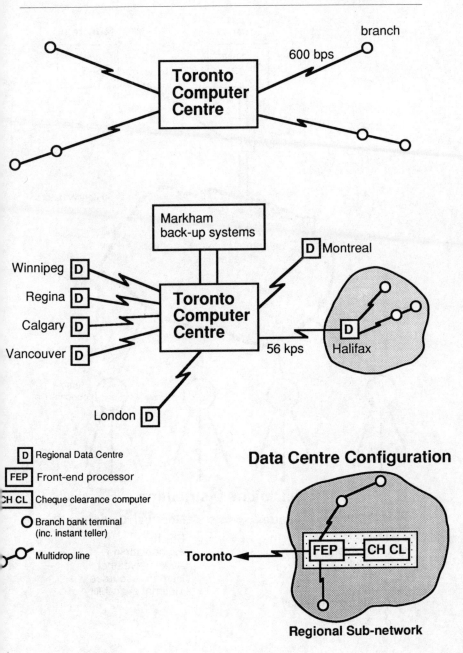

branch

600 bps

Toronto
Computer
Centre

Markham
back-up systems

D Montreal

Winnipeg D

Regina D

Calgary D

Vancouver D

Toronto
Computer
Centre

56 kps

Halifax

D

London D

D Regional Data Centre

FEP Front-end processor

CH CL Cheque clearance computer

O Branch bank terminal
(inc. instant teller)

Multidrop line

Data Centre Configuration

FEP CH CL

Toronto

Regional Sub-network

The Canadian Imperial Bank of Commerce Network

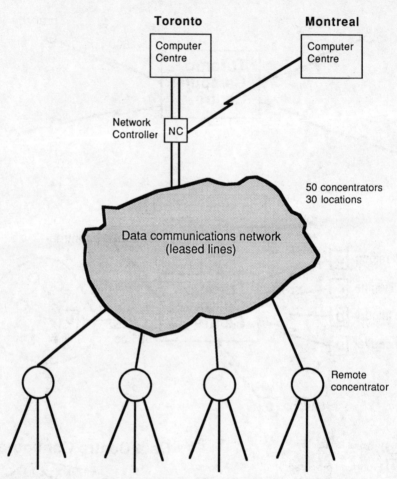

Toronto

Computer
Centre

Montreal

Computer
Centre

Network
Controller   NC

50 concentrators
30 locations

Data communications network
(leased lines)

Remote
concentrator

Applications Distribution

Toronto Computer Centre

Reservations
Flight movement
Pricing/ticketing
Seat selection
Load planning
Flight information display

Montreal Computer Centre

Batch
Flight planning
Aviation weather
Human resources
Material availability

The Air Canada Network

References

Abler, R., 1974, 'The geography of communications', in *Transportation Geography*, M. Eliot Hurst (ed.), New York: McGraw-Hill, pp. 327–46.

——— 1975, 'Effects of space adjusting technologies on the human geography of the future', in R. Abler, D. Jannelle, A. Philbrick and J. Sommer (eds), *Human Geography in a Shrinking World*, North Scituate, MA: Duxburg Press.

——— 1977, 'The telephone and the evolution of the American metropolitan system', in *The Social Impact of the Telephone*, I. de Sola Pool (ed), op.cit.

——— 1987, 'What if somebdoy built a teleport and nobody called? Communications and regional development in the United States', Paper presented at the Seminar on Information and Telecommunications Technology for Regional Development, Government of Greece, OECD, Athens, 7–9 December.

Aglietta, M., 1979, *A Theory of Capitalist Regulation: The US Experience*, London: New Left Books.

Alchian, A. and Demsetz, H., 1972, 'Production, information costs and economic organisation', *American Economic Review*, December, 777–95.

Anderson, A., 1986, 'Presidential address: the four logistical revolutions, *Papers of the Regional Science Association*, 59, pp. 1–12.

Andreff, W., 1984, 'The international centralisation of capital and the reordering of world capitalism', *Capital and Class*, 22, pp. 58–84.

Antonelli, C., 1981, 'Transborder data flows and international business — a pilot study'. Background paper, Committee on Information, Computer and Communications Policy, OECD, Paris.

——— 1988a, 'A new industrial organisation approach', in C. Antonelli (ed), *New Information Technology and Industrial Change: The Italian Case*, Dordrecht-London: Kluwer Academic, pp. 1–12.

——— 1988b, 'The emergence of the network firm', in C. Antonelli (ed), *New Information Technology and Industrial Change: The Italian Case*, Dordrecht-London: Kluwer Academic, pp. 13–32.

Aoki, M., 1984, 'Innovative adaptation through the quasi-tree structure: An emerging aspect of Japanese entrepreneurship', *Zeitschrift für National Ökonomie*, 4, pp. 25–35.

Armstrong, H. and Taylor, T, 1985, *Regional Economics and Policy*, Oxford: Philip Allen.

Arrow, K., 1969, 'The organisation of economic activity', *The Analysis and Evaluation of Public Expenditure: The PPB System*, Joint Economic Committee, 91st Congress 1st Session, 1969, pp. 59–73.

—— 1979, 'The economics of information', in *The Computer Age: A Twenty Year View*, M. Dertouzos and J. Moses (eds), Cambridge, MA: MIT Press.

—— 1984, *The Economics of Information: Collected Papers of K.J. Arrow*, Cambridge, MA: The Belknap Press of Harvard University.

—— 1985, 'Informational structure of the firm', *American Economic Review*, May, pp. 303–7.

Ayres, R., 1984, *The Next Industrial Revolution*, Cambridge, MA: Ballinger.

Baddeley, S. and Dawes, N., 1987, 'Information technology support for devolution', *Local Government Studies*, July/August, pp. 1–16.

Bakis, H., 1982, 'The geographical impact of telecommunications systems used within firms', paper presented at Meeting of the Commission on Industrial Systems, Industrial Geographical Union, University of Sao Paulo, Brazil, August 5–15.

—— 1987, 'Telecommunications in the global firm', in *Industrial Change in Advanced Economies*, F. Hamilton (ed.), London: Croom Helm, pp. 130–60.

Baldwin, C., 1986, 'The capital factor: the impact of home and host countries on the global corporation's cost of capital', in M. Porter (ed.), *Competition in Global Industries*, Cambridge, MA: Harvard Business School Press.

Bank for International Settlements, 1986, *Recent Innovations in International Banking*, Basle: BIS.

Bank of England, 1988, 'The equity market crash', *Quarterly Bulletin*, 28 (1) 51–8.

Baran, B., 1985, 'Office automation and women's work. The technological transformation of the insurance industry', in M. Castells (ed.), *High Technology, Space and Society*, Beverly Hills: Sage.

Barzel, Y., 1982, 'Measurement cost and the organisation of markets', *Journal of Law and Economics*, 25, pp. 27–48.

Baumol, W., 1962, 'On the theory of the expansion of the firm', *American Economic Review*, 5 2, pp. 1078–87.

Bell, D., 1973, *The Coming of Post-Industrial Society*, New York: Basic Books.

—— 1979, 'The social framework of the information society', in M. Dertouzos and J. Moses (eds), *The Computer Age: A Twenty-year View*, Cambridge, MA: MIT Press.

—— and Hart, R., 1980, 'The Regional Demand for Labour Services',

Scottish Journal of Political Economy, 27, pp. 140–51.

Beniger, J., 1986, *The Control Revolution: Technological and Economic Origins of the Information Society*, Cambridge, MA: Harvard University Press.

Benson, J., 1975, 'The interorganisational network as a political economy', *Administrative Science Quarterly*, June pp. 229–49.

Bergstrom, G., Koeneman, J. and Siegel, M., 1983, 'International securities market', in F. Fabozzi (ed.), *Readings in Investment Management*, Homewood, Ill: Richard Irwin.

Berle, A. and Means, G., 1932, *The Modern Corporation and Private Property*, New York: Commerce Clearing House.

Berry, B., 1970, 'The geography of the US in the year 2000', *Transactions of the Institute of British Geographers*, 51, pp. 21–53.

Blair, R. and Kaserman, D., 1983, *Law and Economics of Vertical Integration and Control*, New York: Academic Press.

Blazar, W., Spector, M. and Grathwol, J., 1985, 'The sky above, the teleport below', *Planning*, December, pp. 22–6.

Bliss, C., 1987, 'Arrow's vision of the economic process', in G. Feivel (ed.), *Arrow and the Ascent of Modern Economic Theory*, London: Macmillan, pp. 295–305.

Blois, K., 1972, 'Vertical quasi-integration', *Journal of Industrial Economics*, 20, pp. 253–72.

Bluestone, B., 1987, 'Coping with labour and community: capitalist strategies in the 1980s', in H. Muegge, W. Stohr, P. Hesp and B. Stuckey (eds), *International Economic Restructuring and the Regional Community*, Avebury: Gower.

—— and Harrison, B., 1982, *The Deindustrialisation of America*, New York: Basic Books.

Borchert, J., 1978, 'Major control points in American economic geography', *Annals of the Association of American Geographers*, LXIII, 68, 214–32.

Botts, H. and Patterson, J., 1987, 'Pension fund investments: an initial geographical assessment', *Professional Geographer*, 39(4), 416–27.

Boulding, K., 1963, 'The Knowledge Industry', *Challenge*, May, pp. 36–8.

Britton, J. and Gilmour, J., 1978, *The Weakest Link: A Technological Perspective on Canadian Industrial Underdevelopment*, Background Study 43, Science Council of Canada, Ottawa.

Brotchie, J., Hall, P. and Newton, P. (eds), 1987, *The Spatial Impact of Technological Change*, London: Croom Helm.

Brotchie, J., Newton, P., Hall, P. and Nijkamp, P. (eds), 1985, *The Future of Urban Form: The Impact of New Technology*, London: Croom Helm.

Bruce, M. 1988, 'Home interactive telematics: new technology with a history', in F. van Rijn and R. Williams (eds), *Concerning Home Telematics*, Amsterdam: North Holland.

Bruce, R., Cunard, J. and Director, M. 1987, *The Telecommunications Mosaic*, Frome, Somerset: Butterworths.

Brusco, A., 1982, 'The Emilian model: productive decentralisation and social integration', *Cambridge Journal of Economics*, 6, pp. 167–84.

Camagni, R., 1987, 'The spatial implications of technological diffusion and economic restructuring in Europe', Group on Urban Affairs Meeting on 'Technological Developments and Urban Change', OECD, Paris.

—— 1988, 'Functional integration and locational shifts in new technology industry', in P. Aydalot and D. Keeble (eds), *High Technology Industry and Innovative Environments: The European Experience*, London: Routledge.

—— and Rabellotti, R. (1988), 'Technology, innovation and industrial structure in the textile industry in Italy', Paper presented at the conference on The Application of New Technologies in Existing Industries, Newcastle University, 23–5 March.

Canada Department of Communications, 1987, 'Impact of Telecommunications on Regional Economic Development: Survey of Canadian Organisations on Their Use and the Importance of Telecommunications', Report prepared for the Government of Canada by The Coopers and Lybrand Consulting Group, Ottawa: Canada Department of Communications.

Casson, M., 1987, *The Firm and the Market*, Cambridge, MA: MIT Press.

Castells, M., 1985, 'High technology, economic restructuring, and the urban-regional process in the United States', in M. Castells (ed), *High Technology, Space and Society*, Beverly Hills: Sage.

—— 1987, 'Technological change, economic restructuring and the spatial division of labour', H. Muegge, W. Stohr, P. Hesp and B. Stuckey (eds), *International Economic Restructuring and the Regional Community*, Aldershot: Avebury – Gower.

Caves, R., 1982, *Multinational Enterprise and Economic Analysis*, Cambridge: Cambridge University Press.

Chandler, A., 1962, *Strategy and Structure: Chapters in the History of Industrial Enterprise*, Cambridge, MA: MIT Press.

—— 1977, *The Visible Hand: The Managerial Revolution in American Business*.

Charles, D., 1988, 'Globalisation of R & D and Spatial Variations in Access to Technical Information', paper presented at the ESRC Urban and Regional Seminar Group Meeting, University of Newcastle upon Tyne, 6–8 July.

Cheung, S., 1983, 'The contractual nature of the firm', *Journal of Law and Economics*, XXVI (April), pp. 1–21.

Ciborra, C., 1983, 'Markets bureaucracies and groups in information society', *Information Economics and Policy*, 1, pp. 145–60.

Clark, C., 1940, *The Conditions of Economic Progress*, London: Macmillan.

Clark, G., Gertler, M. and Whiteman, J., 1986, *Regional Dynamics: Studies in*

Adjustment Theory, Boston: Allen & Unwin.

Coakley, J. and Harris, L., 1983, *The City of Capital*, Oxford: Basil Blackwell.

Coase, R., 1937, 'The nature of the firm', *Economica*, 4, pp. 386–405, reprinted in G. Stigler and K. Boulding (eds), *Readings in Price Theory*, Homewood, Ill: Richard D. Irwin, 1952, pp. 331–51.

—— 1960, 'The problem of social cost', *Journal of Law and Economics*, 3, October, pp. 1–44.

Coffey, W. and Polese, M. (eds) 1987, *Still Living Together: Recent Trends and Future Directions in Canadian Regional Development*, Montreal: Institute for Research on Public Policy.

Cohen, S. and Zysman, J., 1987, *Manufacturing Matters*, New York: Basic Books.

Cooper, C., 1983, 'The Structure and Future of the Information Economy', *Information Processing and Management*, 19 (1) pp. 9–26.

—— 1984, 'Is there a need for reform?', in *The International Monetary System: Forty Years After Bretton Woods*, Federal Reserve Bank of Boston, Conference Series, No. 28, pp. 21–39.

Cooper, R., 1972, 'Towards an international capital market', in J. Dunning (ed.), *International Investment*, Harmondsworth: Penguin.

Corey, K., 1982, 'Transactional forces and the metropolis', *Ekistics*, 44, pp. 416–23.

—— 1987, 'Planning the information age metropolis: the case of Singapore', in L. Guclke and R. Preston (eds), *Abstract Thoughts: Concrete Solutions*, University of Waterloo, Department of Geography Publication Series, No. 29.

Cowling, K., 1984, 'The internationalisation of production and de-industrialisation', *University of Warwick*, mimeo.

Cruise O'Brien, R. and Helleiner, G., 1983, 'The political economy of information in a changing international economic order', in R. Cruise O'Brien (ed), *Information, Economics and Power*, London: Hodder and Stoughton.

Cukierman, A., 1984, *Inflation, Stagflation, Relative Prices and Information*, Cambridge: Cambridge University Press.

Cyert, R. and March, J., 1963, *A Behavioural Theory of the Firm*, Englewood Cliffs, NJ: Prentice-Hall.

Dahlman, C., 1979, 'The problem of externality', *Journal of Law and Economics*, 22, pp. 141–62.

Daniels, P., 1985, 'The geography of services', *Progress in Human Geography*, 9, pp. 443–51.

—— 1986, 'Producer services and the post-industrial space economy', in R. Martin and R. Rowthorn (eds), *The Geography of Deindustrialisation*, London: Macmillan.

—— 1987, 'The geography of services' *Progress in Human Geography*, 11(3), 433–47.

—— 1988, 'Some perspectives on the geography of services', *Progress in Human Geography*, 12(3), 431–40.

Davies, S., 1987, 'Vertical integration', in R. Clarke and T. McGuinness (eds), *The Economics of the Firm*, Oxford: Basil Blackwell.

De Meza, D. and van der Ploeg, F., 1987, 'Production flexibility as a motive for multinationality', *The Journal of Industrial Economics*, 35, pp. 343–51.

Demac, D., 1988, 'Hearts and minds revisited: the information policies of the Reagan administration', in V. Mosco and J. Wasko (eds), *The Political Economy of Information*, University of Wisconsin Press.

Dini, L., 1984, 'Towards a European integrated financial market', *Banca Nazionale del Lavoro*, 153, pp. 377–89.

Dobilas, G., 1988, 'Technology and simultaneous financial markets', Ph.D. in progress, Department of Geography, London School of Economics, mimeo.

Dordick, H., Nanus, B. and Bradley, H., 1981, *The Emerging Network Marketplace*, Norwood, NJ: Ablex.

Drucker, P., 1988, 'The coming of the new organisation', *Harvard Business Review*, Jan/Feb, pp. 45–53.

Dunning, J., 1977, 'Trade location of economic activity and the MNE: a search for an eclectic approach', in B. Ohlin, P. Hesselborn and P. Wijkman (eds), *The International Allocation of Economic Activity*, London: Macmillan.

—— and Norman, G., 1987, 'The location choice of offices in international companies', *Environment and Planning A* 19, pp. 613–31.

Dutton, W., Blumler, J. and Kraemer, K. (eds), 1987, *Wired Cities*, Boston, MA: G. K. Hall.

Ebel, K-H., 1985, 'Social and labour implications of flexible manufacturing systems', *International Labour Review*, 124, pp. 133–45.

Economist Publications, The, 1986, *Tokyo 2000: The World's Third International Financial Centre*, Special Report No. 1055, Economist Advisory Group, London: The Economist Publications.

Elcock, H., 1986, 'Decentralisation as a tool for social services management', *Local Government Studies*, July/August, pp. 35–49.

Elger, T., 1987, 'Flexible futures? New technology and the contemporary transformation of work', *Work, Employment and Society*, 1, 528–40.

Espeli, T. and Godo, H., 1986, 'Transborder data flows: a detour towards future information society?', *Norwegian Telecommunications Administration*, Research Department, Report No. 52/86, September.

Estall, R., 1985, 'Stock control in manufacturing: the just-in-time system and its locational implications', *Area*, 17, pp. 129–33.

Estrin, D., 1985, 'Inter-organisational networks: stringing wires across administrative boundaries', *Computer Networks and ISDN Systems*, 9, pp. 281–95.

—— 1986, 'Access to Inter-Organisation Computer Networks', unpublished Ph.D. thesis, Department of Electrical Engineering and Computer Science, MIT, Cambridge, MA.

Eun, C. and Resnick, B., 1988, 'Exchange rate uncertainty, forward contracts, and international portfolio selection', *Journal of Finance*, XLIII (1), 197–215.

Feenberg, A., 1980, 'The political economy of social space', in K. Woodward (ed), *The Myths of Information: Technology and Postindustrial Culture*, London: Routlege & Kegan Paul.

Flaherty, M., 1981, 'Prices versus quantities and vertical financial integration', *Bell Journal of Economics*, 12, pp. 507–25.

Foote, N. and Hart, P., 1953, 'Social mobility and economic advancement', *American Economic Review*, May, pp. 364–77.

Frazier, G., Spekman, R. and O'Neal, C., 1988, 'Just-in-time exchange relationships in industrial markets', *Journal of Marketing*, 52, pp. 52–67.

Freeman, C., 1987, *Technology, Policy and Economic Performance*, London: Frances Pinter.

—— 1988, 'The factory of the future: the productivity paradox, Japanese just-in-time and information technology', *Programme on Information and Communication Technology*, Working Paper 3, London: ESRC.

French, K. and Roll, R., 1986, 'Stock return vacancies', *Journal of Financial Economics*, 17 (1) 5–26.

Friedmann, J. and Wolff, G., 1982, 'World city formation', *International Journal of Urban and Regional Research*, 6, pp. 307–44.

Gandy, O., 1988, 'The political economy of communications competence', in V. Mosco and J. Wasko (eds), *The Political Economy of Information*, University of Wisconsin Press.

Garbade, K. and Silber, W., 1978, 'Technology, communication and the performance of financial markets: 1840–1975', *Journal of Finance*, 33 (3) 819–32.

Gee, K., 1982, *Local Area Networks*, The National Computing Centre Limited, UK.

Gershuny, J., 1978, *After Industrial Society: The Emerging Self-service Economy*, London: Macmillan.

—— and Miles, I., 1983, *The New Service Economy*, London: Frances Pinter.

Gertler, M., 1984, 'Regional capital theory', *Progress in Human Geography*, 8, pp. 50–81.

—— 1986a, 'Some problems of time in economic geography', paper presented at the Annual Meeting of the Association of American

Geographers, Minneapolis MN, 6 May.

—— 1986b, 'Regional dynamics of manufacturing and non-manufacturing investment in Canada', *Regional Studies*, 20 (6), 523–34.

—— 1988, 'The limits to flexibility: comments on the post-Fordist vision of production and its geography', *Transactions of the Institute of British Geographers*, 13, pp. 419–32.

Gillespie, A., 1987, 'Telecommunications and the development of Europe's less-favoured regions', *Geoforum*, 18(2), 229–36.

—— Goddard, J., Robinson, F., Smith, I. and Thwaites, A., 1984, *The Effects of New Information Technology in the Less-Favoured Regions of the Community*, Brussels: EEC.

—— and Hepworth, M., 1988, 'Telecommunications and regional development in the information economy', *ESRC Programme on Information and Communication Technologies*, Working Paper 1, January, London: ESRC.

—— and Williams, H., 1988, 'The impacts of telecommunications liberalisation on small and medium-sized enterprises', Paper submitted to ICCP/ OECD, Paris, March.

Glasmeier, A., 1988, 'The Japanese technopolis programme: high-tech development strategy or industrial policy in disguise?', *International Journal of Urban and Regional Research*, 12(2) pp. 268–83.

Gleed, R. and Rees, R., 1979, 'The derivation of regional capital stock estimates for UK manufacturing industries 1951–73', *Journal of the Royal Statistical Society*, 142, pp. 330–46.

Goddard, J., 1975, *Office Location in Urban and Regional Development*, London: Oxford University Press.

—— 1980, 'Technological forecasting in a spatial context', *Futures*, 12(2), 90–105.

—— and Gillespie, A., 1986, 'Advanced telecommunications and regional economic development', *The Geographical Journal*, 152, pp. 383–97.

—— Gillespie, A., Thwaites, A. and Robinson, F., 1986, 'The impact of new information technology on urban and regional structure in Europe', *Land Development Studies*, 3, pp. 11–32.

—— and Pye, R., 1977 'Telecommunications and office location', *Regional Studies*, 11, pp. 19–30.

Goldberg, V., 1985, 'Production functions, transaction costs and the new institutionalism', in G. Feiwel (ed.), *Issues in Contemporary Microeconomics and Welfare*, London: Macmillan.

Golding, P. and Murdock, G., 1986, 'Unequal information: access and exclusion in the new communications marketplace', in M. Ferguson (ed.), *New Communications Technology and the Public Interest*, London: Sage.

Goldmark, P., 1972, 'Communication and the Community', in *Communication, A Scientific American Book*, San Francisco: W. H. Freeman.

Gordon, R. and Kimball, L., 1987, 'The impact of industrial structure on global high technology location', in J. Brotchie *et al.* (eds), *The Spatial Impact of Technological Change*, pp. 157–84.

Gottmann, J., 1961, *Megalopolis: The Urbanised Northeastern Seaboard of the United States*, New York: The Twentieth Century Fund.

—— 1983, *The Coming of the Transactional City*, Institute for Urban Studies, University of Maryland, Monograph Series, Number 2.

Graham, M. and Rosenthal, S., 1986, 'Flexible manufacturing systems require flexible people', *Human Systems Management*, 6, pp. 211–22.

Graham, S., 1988, 'Conceptualising the role and nature of information in contemporary local government', Department of Town and Country Planning, University of Newcastle upon Tyne, mimeo.

Greenspan, A., 1988, 'Statements to Congress', *Federal Reserve Bulletin*, 74 (2), 91–105.

Grimshaw, D. and Haddad, A., 1988, 'Trends in the use of information technology in local government', *Local Government Studies*, July/August, pp. 15–25.

Grossman, S., 1976, 'On the efficiency of competitive stock markets where traders have diverse information', *Journal of Finance*, pp. 573–85.

Hakim, C., 1987, 'Trends in the flexible workforce', *Employment Gazette* No. 95, pp. 549–60.

Hall, P., 1987, 'The anatomy of job creation: rations, regions and cities in the 1960s and 1970s, *Regional Studies*, 21, pp. 95–106.

—— 1989, *London 2001*, London: Unwin Hyman.

—— and Preston, P., 1988, *The Carrier Wave: New Information Technology and the Geography of Innovation, 1846–2003*, London: Unwin Hyman.

Hall, R. and Hitch, C., 1939, 'Price theory and business behaviour', *Oxford Economic Papers*, 2, pp. 12–45.

Hambleton, R., 1988, 'Consumerism, decentralisation and local democracy', *Public Administration*, 66, pp. 125–47.

Hamelink, C., 1984, *Transnational Data Flows in the Information Age*, Lund: Gleerup.

Hamilton, A., 1986, *The Financial Revolution*, Harmondsworth: Viking/ Penguin.

Hamilton, J., 1978, 'Marketplace organisation and marketability: NASDAQ, The Stock Exchange and the national market system', *Journal of Finance*, 33 (2), 487–503.

Harvey, D., 1987, 'Flexible accumulation through urbanisation: reflections on post-modernism in the American city', *Antipode*, 19, pp. 260–86.

—— and Scott, A., 1988, 'The practice of human geography: theory and empirical specificity in the transition from Fordism to flexible accumulation', in W. Macmillan (ed.) *Remodelling Geography*, Oxford:

Basil Blackwell.

Helleiner, G, 1983, 'Uncertainty, information and the economic interests of developing countries', in R. Cruise O'Brien (ed.), *Information, Economics and Power*, London: Hodder and Stoughton.

Hepworth, M., 1986a, 'The geography of technological change in the information economy', *Regional Studies*, 20, pp. 407–24.

—— 1986b, 'The geography of economic opportunity in the information society', *The Information Society*, 4(3), 205–20.

—— 1987a, 'The geography of the information economy' unpublished Ph.D., Department of Geography, University of Toronto.

—— 1987b, 'Information technology as spatial systems, *Progress in Human Geography*, 11, pp. 157–80.

—— 1987c, 'The information city' *Cities*, August, pp. 253–62.

—— 1988a, 'Information technology and the global restructuring of capital markets', in T. Leinbach and S. Brunn (eds), *Collapsing Space and Time: Geographical Perspectives on Communication and Information*, New York: Allen & Unwin (forthcoming).

—— 1988b, 'Planning for the information city: the challenge and response', *Newcastle Studies of the Information Economy*, Working Paper, No. 6., CURDS, University of Newcastle upon Tyne.

—— 1989, 'The electronic comeback of information space', *Tijdschrift voor Economische en Sociale Geografie*, 30 (forthcoming).

—— 1989b, 'Geographical advantage in the information economy, in E. Punset and G. Sweeney (eds), *Information Resources and Corporate Growth*, London: Pinter.

—— and Dobilas, G., 1985, 'The city and the information revolution', *Urban Resources*, 3, Fall, pp. 39–46.

—— Green, A. and Gillespie, A., 1987, 'The spatial division of information labour in Great Britain', *Environment and Planning A*, 19, pp. 793–806.

—— and Robins, K., 1987, 'Whole information society: a view from the periphery', *Media, Culture and Society*, 10 (3) 323–45.

—— and Waterson, M., 1988, 'Information Technology and the spatial dynamics of capital', *Information Economics and Policy*, Vol 3, pp. 148–63.

Hepworth, N., 'The future of local government finance', *Policy Studies* 8, (3) 1–8.

Hilbig, M. and Monse, K., 1988, 'Home interactive telematics and new services: strategies and trends in restructuring the service sector', in F. van Rijn and R. Williams (eds), *Concerning Home Telematics*.

Hodgson, G., 1988, *Economics and Institutions*, Oxford: Policy Press.

Holland, S., 1976, *Capital Versus the Regions*, London: Macmillan.

Hoover, E.M., 1975, *An Introduction to Regional Economics*, New York: Alfred Knopf.

Howells, J., 1987, 'Developments in the location, technology and industrial organisation of computer services: some trends and research issues', *Regional Studies*, 21, 6, pp. 493–503.

—— 1988, *Economic, Technological and Locational Trends in European Services*, Aldershot: Avebury.

—— 1989, 'Location and organisation of R & D: new horizons', *Research Policy*, 18 (forthcoming).

Hudson, E.A. and Jorgenson, K., 1974, 'US energy policy and economic growth, 1975–2000', *Bell Journal of Economics*, 5, p. 461.

Huet, A., 1988, 'Téléports et services à valeur ajouté', Meeting of the European Study Group of the International Geographical Union on 'Telecommunications and Territory', Paris.

Ihda, K., 1988, 'Osaka teleport as the hub of the Technoport Osaka Development', in Noothoven van Goor and Lefcoe (eds), 1987, op. cit.

Information Technology Advisory Panel (ITAP), 1983, *Making a Business of Information*, UK Cabinet Office, London: HMSO.

Irwin, M., 1988, 'Corporate strategy and information networks', *Intermedia*, 16(2), 32.–6.

Itoh, S., 1987, 'Teleports and their impact on world economy', in Noothoven van Goor and Lefcoe (eds), 1987, op. cit.

Jacquemin, A., 1987, *The New Industrial Organisation*, Cambridge, MA: MIT Press.

Jaffe, J. and Rubinstein, M., 1987, 'The value of information in impersonal and personal markets', in T. Copeland (ed.), *Modern Finance and Industrial Economics*, Oxford: Basil Blackwell.

Johansson, B., 1987, 'Information technology and the viability of spatial networks', *Papers of the Regional Science Association*, 61, pp. 51–64.

Johanson, J. and Vahlne, J., 1977, 'The internationalisation process of the firm: a model of knowledge development and increasing foreign market commitments', *Journal of International Business Studies*, 8, pp. 23–32.

Johnson, S., 1988, 'The changing division of information labour in the UK', PICT Working Paper, MM2, CURDS, University of Newcastle upon Tyne, mimeo.

Jones, B. and Scott, P., 1987, 'Flexible manufacturing systems in Britain and the USA', *New Technology, Work and Employment*, 2, pp. 27–36.

Jonscher, C., 1983, 'Information resources and economic productivity', *Information Economics and Policy*, 2, (1) 13–35.

Karunaratne, N., 1986, 'Issues in measuring the information economy', *Journal of Economic Studies*, 13, pp. 51–65.

Keen, P., 1986, *Competing in Time*, Cambridge, MA: Ballinger.

Kellerman, A., 1984, 'Telecommunications and the geography of metropolitan areas', *Progress in Human Geography*, 8, pp. 222–46.

Kincaid, G., 1988, 'Policy implications of structural changes in financial markets', *Finance and Development*, March, pp. 2–5.

Kindleberger, C., 1983, 'Key currencies and financial centres', in F. Machlup G. Fels and H. Muller-Groeling (eds), *Reflections on a Troubled World Economy*, London: Macmillan.

—— 1987, *International Capital Movements*, Cambridge University Press.

King, D. and Gurr, T., 1988, 'The state and fiscal crisis in advanced industrial democracies', *International Journal of Urban and Regional Research*, 12, pp. 87–105.

Klein, B., Crawford, R. and Alchian, A., 1978, 'Vertical integration, appropriable rents and the competitive contracting process', *Journal of Law and Economics*, 21(2), 297–326.

Klein, B. and Leffler, K., 1981, 'The role of market forces in assuring contractual performances', *Journal of Political Economy*, 89, pp. 615–41.

Kling, R., 1988, 'Building an institutionalist theory of regulation', *Journal of Economic Issues*, 22, March, pp. 197–209.

Knight, J., 1984, 'The interconnection of European stock exchanges', *The Stock Exchange Quarterly*, December, pp. 14–16.

Kohlhagen, S., 1983, 'Overlapping national investment portfolios', *Research in International Business and Finance*, 3, London: JAI Press, pp. 113–37.

Korte, W., Robinson, S. and Steinle, W., 1987, *Telework: Present Situation and Future Development of a New Form of Work Organisation*, Amsterdam: North Holland.

Kraemer, K. and King, J., 1988, 'The role of information technology in managing cities', *Local Government Studies*, March/April, pp. 23–47.

Kutay, A., 1986a, 'Optimum office location and the comparative statics of information economies', *Regional Studies*, 20 (6), 551–64.

—— 1986b, 'Effects of telecommunications on office location', *Urban Geography*, 7, pp. 243–57.

—— 1988, 'Technological change and spatial transformation in the information economy: 2. The influence of new information technology on the urban system', *Environment and Planning A*, 20, pp. 707–18.

Lachmann, L., 1978, *Capital and Its Structure*, Kansas City: Sheed Andrews and McMeel.

Laffont, J-J., 1987, 'Toward a normative theory of incentive contracts between government and private firms', *The Economic Journal*, 97, pp. 17–31.

Lamberton, D., 1983, 'Information economics and technological change', in S. MacDonald, D. Lamberton and T. Mandeville (eds), *The Trouble with Technology: Explorations in the Process of Technological Change*, London: Frances Pinter.

Lambooy, J., 1986, 'Information and internationalisation: dynamics of the relations of small and medium sized enterprises in a network environment', paper presented at the *Round Table on Les PME innovatrices et leur environnement local et economique*, Aix-en-Provence, July.

Langdale, J., 1983, 'Competition in the United States long-distance telecommunications industry', *Regional Studies*, 17, pp. 393–409.

—— 1985, 'Electronic funds transfer and the internationalisation of the banking and finance industry', *Geoforum*, 16 (1) 1–13.

—— 1987, 'Telecommunications and electronic information services in Australia', in J. Brotchie, P. Hall, P. Newton (eds), *The Spatial Impact of Technological Change*, London: Croom Helm, pp. 89–103.

Lange, S., 1987, 'The philosophy of teleports in Germany developed by Fraunhofer-ISI Karlsruhe', in J. Noothaven van Goor and G. Lefcoe (eds), op cit.

Lazerson, M., 1988, 'Organisational growth of small firms: an outcome of markets and hierarchies?', *American Sociological Review*, 53 (June), pp. 330–42.

Lemoine, P., 1981, 'Information and economic development', in *Information Activities and The Role of Electronics and Telecommunications Technologies*, OECD, ICCP Series No. 6, Volume 2, Group of Experts Studies, OECD, Paris.

Lessard, D., 1988, 'Finance and global competition: exploiting financial scope and coping with volatile exchange rates', in J. Stern and D. Chew (eds), *New Developments in International Finance*, New York: Basil Blackwell.

Lesser, B., 1987. 'Telecommunications services and regional development in Canada: the case of the Atlantic Provinces', paper presented at the OECD seminar on Information and Telecommunications Technology for Regional Development, Athens, December 7–9.

Leyshon A., Thrift, N. and Daniels, P., 1987, 'the urban and regional consequences of the restructuring of world financial markets: the case of the City of London', Working Paper on Producer Services, No. 4, Departments of Geography, University of Bristol/Liverpool.

Loasby, B., 1976, *Choice, Complexity and Ignorance*, Cambridge University Press.

Loinger, G. and Peyrache, V., 1988, 'Technological clusters and regional economic restructuring' in P. Aydalot and D. Keeble (eds), *High Technology Industry and Innovative Environments: The European Experience*, London: Routledge.

Lowi, T., 1980, 'The political impact of information technology' in T. Forester (ed.), *The Microelectronics Revolution*, Oxford: Blackwell.

McCall, J., Noll, R. and Spence, M., 1983, 'Information economics and policy in America', *Information Economics and Policy*, 1, pp. 11–12.

McGuinness, T., 1987, 'Markets and managerial hierarchies', in R. Clarke and T. McGuinness (eds), *The Economics of the Firm*, Oxford: Basil Blackwell.

Machlup, F., 1962, *The Production and Distribution of Knowledge in the United States*, Princeton, NJ: Princeton University Press.

Macpherson, A., 1988, 'Service-to-manufacturing linkages and industrial innovation: empirical evidence from metropolitan Toronto', unpublished Ph.D. thesis, Department of Geography, University of Toronto.

Maier, G., 1985, 'Cumulative causation and selectivity in labour market oriented migration caused by imperfect information', *Regional Studies* 19 (3), 231–41.

Malmgren, H., 1961, 'Information, expectations and the theory of the firm', *Quarterly Journal of Economics*, 75, 339–421.

Markusen, J., 1984, 'Multinationals, multi-plant economies and the gains from trade', *Journal of International Economics*, 16, pp. 205–26.

Marris, R., 1964, *The Economic Theory of Managerial Capitalism*, London: Macmillan.

Marschak, J., 1968, 'Economics of inquiring, communicating and deciding', *American Economic Review*, 58(2), pp. 1–18.

Marshall, J., 1988, 'Corporate reorganisation and the geography of services', paper presented at the ESRC Urban and Regional Seminar Group Meeting, University of Newcastle upon Tyne, 6–8 July.

Martin, J., 1978, *The Wired Society*, Englewood Cliffs, NJ: Prentice-Hall.

—— 1981, *Computer Networks and Distributed Processing*, Englewood Cliffs: Prentice-Hall.

Martin, R., 1987, 'The new economics and politics of regional restructuring: the British experience'. Paper presented at the International Conference on *Regional Policy at the Crossroads*, University of Leuven, Belgium, April 22–4.

Massey, D., 1984, *Spatial Divisions of Labour*, London: Macmillan.

Meier, R., 1962, *A Communications Theory of Urban Growth*, Cambridge, MA: MIT Press.

Meijer, A. and Peeters, P., 1982, *Computer Network Architectures*, London: Pitman.

Melody, W., 1986, 'Telecommunications — policy directions for the technology and information services', *Oxford Surveys in Information Technology*, Vol. 3.

—— 1987, 'Information: an emerging dimension of institutional analysis', Invited Paper for the Paradigm Papers Project, *Journal of Economic Issues* (forthcoming).

Mettler-Meibom, B., 1988, 'Communication at Stake?' in *Concerning Home Telematics*, F. van Rijn and R. Williams (eds), op. cit.

Meyer, P., 1986, 'Computer-controlled manufacture and the spatial distribution of production', *Area*, 18, pp. 209–13.

Miles, I., 1988a, 'Information technology and information society: options for the future', Programme on Information and Communication Technologies, PICT Policy Research Paper, No. 2, London: ESRC.

—— 1988b, *Home Informatics*, London: Frances Pinter.

Mills, D., 1984, 'Demand fluctuations and endogenous firm flexibility', *Journal of Industrial Economics*, 33, (1) pp. 55–71.

Milne, S., 1989, 'New forms of manufacturing and their spatial implications: the UK electronic consumer goods industry', *Environment and Planning A* (forthcoming).

Minnerop, H. and Stoll, H., 1986, 'Technological change in the back office: implications for structure and regulation of the securities industry', in A. Saunders and L. White (eds), *Technology and the Regulation of Financial Markets*, Lexington, MA: D. C. Heath.

Minns, R., 1982, *Take over the City*, London: Pluto Press.

Mitropoulos, M., 1983, 'Public participation, as access, in cable TV in the USA', *Ekistics*, 302, October, pp. 385–92.

Monterverde, K. and Teece, D., 1982, 'Supplier switching costs and vertical integration in the automobile industry', *Bell Journal of Economics*, 13, pp. 206–13.

Moore, C. and Booth, S., 1986, 'Urban policy contradictions: the market versus redistributive approaches', *Policy and Politics*, 14, (3), 361–87.

Moore, S., 1987, 'Contracting out: a painless alternative to the budget cutter's knife', in *Proceedings of the Academy of Political Science*, 36 (3), pp. 60–73.

Morphet, J., 1987, 'Local authority decentralisation: Tower Hamlets goes all the way', *Policy and Politics*, 15, pp. 119–26.

Mosco, V, 1982, *Push-Button Fantasies*, Norwood, NJ: Ablex.

—— 1988, 'Information in the Pay-per Society', in V. Mosco and J. Wasko (eds), *The Political Economy of Information*, Madison: University of Wisconsin Press.

Moss, M., 1986a, 'Telecommunications Policy and World Urban Development'. Paper presented at the Annual Meeting of the International Institute of Communications, Edinburgh, September.

—— 1986b, 'Telecommunications and the future of cities', *Land Development Studies*, 3, pp. 33–44.

—— 1987, 'Telecommunications: shaping the future', paper presented at the Conference on America's New Economic Geography, Washington, DC, April.

—— and Dunau, A., 1987, 'Will the cities lose their back offices?' *Real Estate Review*, 17, (1), 62–75.

Murphy, P., 1988, 'Moving on to the offensive', *The Banker*, March, pp. 33–4.

Nasbeth, L. and Ray, G., 1974, *The Diffusion of New Industrial Processes*, Cambridge: Cambridge University Press.

Nelson, K., 1986, 'Labour demand, labour supply and the suburbanisation of low-wage office work', in A. Scott and M. Storper (eds), *Production, Work, Territory*, Boston, MA: Allen & Unwin.

Nicholas, S., 1986, 'The theory of multinational enterprise as a transaction mode', in P. Hertner and E. Jones (eds), *Multinationals: Theory and History*, Aldershot: Gower.

Nicholson, G., 1989 'A model of how not to regenerate an urban area', *Town and Country Planning*, 58, pp. 52–5.

Nicol, L., 1985, 'Communications technology: economic and spatial impacts', in M. Castells (ed), *High Technology, Space and Society*, Beverley Hills: Sage.

Nijkamp, P. and Mouwen, A., 1987, 'Knowledge centres, information diffusion and regional development', in J. Brotchie, P. Hall and P. Newton (eds), *The Spatial Impact of Technological Change*, London: Croom Helm, pp. 254–70.

Noothoven van Goor, J. and Lefcoe, G. (eds), 1987, *Teleports in the Information Age*, Oxford: North Holland.

Noyelle, T. and Stanback, T., 1983, *The Economic Transformation of American Cities*, Totowa, NJ: Allanheld Osmun.

O'Brien, D. and Swann, D., 1968, *Information Agreements, Competition and Efficiency*, London: Macmillan.

OECD, 1981, *Information Activities, Electronics and Telecommunications Technologies*, ICCP Series, Vol. 1, OECD, Paris.

—— 1986, *Trends in the Information Economy*, Information, Computer and Communication Policy Series, No. 11, OECD, Paris.

—— 1987, 'Ownership linkages in financial services', Committee on Financial Markets, OECD, Paris.

Ohlson, J., 1987, *The Theory of Financial Markets and Information*, London: North-Holland.

Openshaw, S. and Goddard, J, 1987, 'Some implications of the commodification of information and the emerging information economy for applied geographical analysis in the UK', *Environment and Planning A* 19, pp. 23–40.

Panzar, J. and Willig, R., 1981, 'Economies of scope', *American Economic Review*, Vol. 71, no. 2, pp. 268–72.

Perez, C., 1985, 'Microelectronics, long waves and world structural change: new perspectives for developing countries', *World Development*, 17, pp. 441–63.

Pfeffer, J. and Salancik, G., 1978, *The External Control of Organisations: A Resource-Dependence Perspective*, New York: Harper and Row.

Pike, R. and Mosco, V., 1986, 'Canadian consumers and telephone pricing: from luxury to necessity and back again', *Telecommunications Policy* 10(1), 17–32.

Piore, M., 1986, 'Perspectives on labour market flexibility', *Industrial Relations*, 25 (2), 146–66.

—— and Sabel, C., 1984, *The Second Industrial Divide*, New York: Basic Books.

Pool, de Sola I (ed.), 1977, *The Social Impact of the Telephone*, Cambridge: MIT Press.

Porat, M., 1977, *The Information Economy: Definition and Measurement*, Special Publication 77–12(1), Office of Telecommunications, US Department of Commerce, Washington.

—— 1978, 'Policy uses of a macroeconomic model of the information sector and microeconomic production functions', Background Paper, Committee on Information, Computer, Communications Policy, OECD, Paris.

Porter, M., 1985, *Competitive Advantage*, New York: The Fine Press.

Pred, A., 1973, 'The growth and development of systems of cities in advanced economies', in A. Pred and G. Tornqvist (eds), *Systems of Cities and Information Flows*, University of Lund, Lund Studies in Geography Series B, No. 38.

—— 1977, *City Systems in Advanced Economies*, London: Hutchinson.

Putterman, L., 1984, 'On some recent explanations of why capital hires labour', *Economic Inquiry*, 22, pp. 171–87.

Pye, R. and Lauder, G., 1987, 'Regional aid for telecommunications in Europe', *Telecommunications Policy*, June, pp. 99–113.

Rada, J., 1984, 'Development, telecommunications and the emerging service economy', paper presented at the Second World Conference on Transborder Data Flow Policies, Rome, 26–9 June.

Radner, R., 1970, 'Problems in the theory of markets under uncertainty', *American Economic Review*, Vol. 60, pp. 454–60.

Rajan, A. and Pearson, R., 1986, *UK Occupation and Employment Trends in 1990*, London: Butterworths.

Richardson, H., 1969, *Regional Economics*, Weidenfeld & Nicolson: London.

Robertson, D., 1923, *The Control of Industry*, London: Nisbet.

Robins, J., 1987, 'Organisational economics: notes on the use of transaction-cost theory in the study of organisations', *Administrative Science Quarterly*, 32, pp. 68–87.

Robins, K. and Hepworth, M., 1988a, 'Electronic spaces: new technologies and the future of cities', *Futures*, 20 (2), 155–76.

—— 1988b, 'Home interactive telematics and the urbanisation process', in F. van Rijn and R. Williams (eds), *Concerning Home Telematics*.

—— and Webster, F., 1980, 'Information is a social relation', *Information* 8, pp. 30–5.

Robinson, J., 1969, *The Economics of Imperfect Competition*, London: Macmillan.

Roobeek, A., 1987, 'The crisis in Fordism and the rise of a new technological paradigm', *Futures*, April, pp. 129–54.

Roszac, T., 1986, *The Cult of Information*, Cambridge: Lutterworth.

Rugman, A., 1981, *Inside the Multinationals: The Economics of Internal Markets*, London: Croom Helm.

Ruttenberg, R., 1987, 'Impact of demographic and technological change on urban areas', Report to the Group on Urban Affairs, OECD, Paris, June.

Rutterford, J., 1983, *Introduction to Stock Exchange Investment*, London: Macmillan.

Ravan, A. and Pearson, R., 1986, *UK Occupation and Employment Trends in 1990*, London: Butterworths.

Sabel, C., 1989, 'Flexible specialisation and the re-emergence of regional economies', in P. Hirst and J. Zeitlin (eds), *Reversing Industrial Decline?*, Oxford: Berg.

Sarcinelli, M., 1986, 'The EMS and the International Monetary System: towards greater staibility', *Banca Nazionali del Lavoro*, 156, pp. 57–81.

Sauvant, K., 1986a, *Trade and Foreign Direct Investment in Data Services*, Boulder: Westview Press.

—— 1986b, *International Transactions in Services: The Politics of Transborder Data Flows*, Boulder, London: Westview Press.

Savoie, D. (ed), 1986, *The Canadian Economy: A Regional Perspective*, Toronto: Methuen.

Scherer, F., 1980, *Industrial Market Structure and Economic Performance*, Chicago: Rand McNally.

Schiller, D., 1982, *Telematics and Government*, Norwood, NJ: Ablex.

Schiller, H., 1984, *Information and the Crisis Economy*, Norwood, NJ: Ablex.

Schoenberger, E., 1987, 'Technological and organisational change in automobile production: spatial implications', *Regional Studies*, 21, pp. 199–214.

Scott, A., 1983, 'Industrial organisation and the logic of intra-metropolitan location I: theoretical considerations', *Economic Geography*, 59, pp. 233–50.

—— 1986a, 'High technology industry and territorial development: the rise of the Orange County Complex, 1955–1984', *Urban Geography*, 1 (1), 3–45.

—— 1986b, 'Industrial organisation and location: division of labour, the firm and spatial process', *Economic Geography*, 62, pp. 215–31.

—— 1988, 'Flexible production systems and regional development: the rise of new industrial spaces in North America and Western Europe', *International Journal of Urban and Regional Research*, 12 (3), 173–85.

—— and Storper, M., 1987, 'High technology industry and regional development: a theoretical critique and reconstruction', *International Social Science Journal*, 112, pp. 215–32.

Scott, A. and Angel, D., 1988, 'The global assembly-operations of US semiconductor firms: a geographical analysis', *Environment and Planning A*, 20, pp. 1047–67.

Scribner, R., 1986, 'The technological revolution in securities trading: can regulation keep up?', in A. Saunders and L. White (eds), *Technology and the Regulation of Financial Markets*, Lexington, MA: D. C. Heath.

Sellgren, J., 1987, 'Local economic development and local initiatives in the mid-1980s', *Local Government Studies*, November/December, pp. 51–68.

Semple, R., Green, M. and Martz, D., 1985, 'Perspectives on corporate headquarters relocation in the United States', *Urban Geography*, 6, pp. 370–91.

Sharma, R., de Sousa, P. and Ingle, A., 1982, *Network Systems*, New York: Van Nostrand Rienhold.

Silber, W., 1983, 'Recent structural change in the capital markets: the process of financial innovation', *American Economic Review*, 73 (2), 89–95.

Silver, M., 1984, *Enterprise and the Scope of the Firm*, Oxford: Martin Robertson & Co.

Simon, H., 1955, 'A behavioural theory of rational choice', *Quarterly Journal of Economics*, 69, pp 99–118.

Singlemann, J., 1978, *From Agriculture to Services*, Beverley Hills: Sage.

Slack, J. and F. Fejes (eds), 1987, *The Ideology of the Information Age*, Norwood, NJ: Ablex.

Smith, R., 1972, *The Wired Nation: Cable TV: The Electronic Communications Highway*, New York: Harper and Row.

Snow, M., 1988, 'Telecommunications literature: a critical review of the economic, technological and public policy issues', *Telecommunications Policy*, 12, pp. 153–83.

Solomon, B., 1985, 'Regional econometric models', *Progress in Human Geography*, 9, pp. 379–99.

Solow, R., 1957, Technical change and the aggregate production function, *Review of Economics and Statistics*, 39, pp. 312–20.

Spence, M., 1974, 'An economist's view of information', in C. Caudra and A. Luke (eds), *Annual Review of Information Science and Technology*, 9, Washington, DC: American Society for Information and Science.

Stanback, T., Bearse, P., Noyelle, T., and Karasek, R., 1981, *Services: The New Economy*, Totowa, NJ: Allanheld and Osman.

Starr, P., 1987, 'The limits of privatisation', *Proceedings of the Academy of Political Science*, 36 (3), pp. 124–37.

Stearns, P., 1977, 'Is there a postindustrial society?', in L. Estabrook (ed.), *Libraries in Postindustrial Society*, Phoenix, AZ: Oryx Press.

Stigler, G., 1939, 'Production and distribution in the short run', *Journal of Political Economy*, Vol. 47 (June), pp. 305–27.

—— 1951, 'The division of labour is limited by the extent of the market', *Journal of Political Economy*, 59, pp. 185–93.

—— 1961, 'The economics of information', *Journal of Political Economy*, 69, pp. 213–25.

Stiglitz, J., 1982, 'Information in capital markets', in W. Sharpe and C. Cootner (eds), *Financial Economics: Essays in Honour of Paul Cootner*, Englewood Cliffs, NJ: Prentice-Hall.

—— 1985, 'Information and economic analysis, *The Economic Journal Supplement*, 95, pp. 21–41.

Stonham, P., 1987, *Global Stock Market Reports*, Aldershot: Gower.

Stonier, T., 1983, *The Wealth of Information*, London: Methuen.

Storper, M. and Christopherson, S., 1987, 'Flexible specialisation and regional industrial agglomerations: the case of the US motion picture industry', *Annals of the Association of American Geographers*, 77, pp. 104–17.

Storper, N., 1987, 'The new industrial geography', *Urban Geography*, 8, (6), 585–98.

Strassmann, P., 1985, *Information Payoff*, New York: The Free Press.

Strong, N. and Waterson, M., 1987, 'Principles, agents and information', in R. Clarke and T. McGuinness (eds), *The Economics of the Firm*, Oxford: Basil Blackwell.

Sugarman, A., 1985, *High Tech Real Estate*, New York: Dow and Jones-Irwin.

Tanenbaum, A., 1981, *Computer Networks*, Englewood Cliffs, NJ: Prentice-Hall.

Taylor, M., 1987, 'Technological change and the business enterprise', in J. Brotchie, P. Hall and P. Newton (eds), *The Spatial Impact of Technological Change*, London: Croom Helm, pp. 208–27.

Teece, D., 1980, 'Economies of scope and the scope of the enterprise', *Journal of Economic Behaviour and Organisation*, 1, pp. 223–47.

Thorelli, H., 1986, 'Networks: Between markets and hierarchies', *Strategic Management Journal*, 7, pp. 37–51.

Thorngren, B., 1970, 'How do contact systems affect regional development?', *Environment and Planning*, 2, pp. 409–27.

Thrift, N. and Leyshon, A., 1988, 'The gambling propensity: banks, developing country debt exposures and the new international financial system', *Geoforum*, 19 (1), 55–69.

Toffler, A., 1981, *The Third Wave*, London: Pan.

Toone, R. and Jackson, D. (eds), 1987, *The Management of Manufacturing*, Bedford: IFS Publications.

Tornqvist, G., 1968, 'Flows of information and the location of economic activites', *Lund Series in Geography*, Series B3.

—— 1974, 'Flows of information and the location of economic activities', in M. Eliot-Hurst (ed.), *Transportation Geography: Comments and Readings*, New York: McGraw Hill.

Van Rijn, F. and Williams, R. (eds), 1988, *Concerning Home Telematics*, Amsterdam: North Holland.

Vickerman, R., 1984, *Urban Economics*, Philip Allen: Oxford.

Vincent, G. and Peacock, J., 1985, *The Automated Building*, London: The Architectural Press.

Wallis, J. and North, D., 1986, 'Measuring the transaction sector in the American economy, 1870–1970', in S. Engerman and R. Gallman (eds), *Long Term Factors in American Economic Growth, Studies in Income and Wealth*, 51, Chicago: University of Chicago Press.

Warskett, G., 1981, 'The role of information activities in total Canadian manufacturing: separability and substitution', ICCP Series, No. 6, 2, OECD, Paris.

Waterson, M., 1987, 'The economics of information and information technology: theory and spatial implications', in *Newcastle Studies of the Information Economy*, No. 2, Centre for Urban and Regional Development Studies, University of Newcastle upon Tyne.

Weatherall, A., 1988, *Computer Integrated Manufacturing*, London: Butterworths.

Webber, M., 1980, 'A communications strategy for the cities of the 21st century', Working Paper, Berkeley: University of California, Institute of Urban and Regional Development.

Whynes, D., 1987, 'On assessing efficiency in the provision of local authority services', *Local Government Studies* January/February, pp. 53–69.

Williamson, O., 1973, 'Markets and hierarchies: some elementary considerations', *American Economic Review*, 63 (May), pp. 316–25.

—— 1975, *Markets and Hierarchies: Analysis and Antitrust Implications*, New York: The Press.

—— 1985, *The Economic Institutions of Capitalism*, New York: The Free Press.

—— 1986a, 'Vertical integration and related variations in a transaction-cost economics theme', in J. Stiglitz and G. Mathewson (eds), *New Developments in the Analysis of Market Structure*, Cambridge, MA: MIT Press.

—— 1986b, 'Hierarchical control and optimum firm size', in *Economic Organisation*, Brighton: Wheatsheaf: Chapter 3, pp. 32–53.

—— 1986c, *Economic Organisation*, Brighton: Wheatsheaf Books.

—— and Bhargava, N., 1986, 'Assessing and clarifying the internal structure and control apparatus of the modern corporation', in *Economic Organisation*, Brighton: Wheatsheaf, Chapter 4, pp. 54–80.

—— Wachter, M. and Harris, J., 1975, 'Understanding the employment relation: the analysis of idiosyncratic exchange', *Bell Journal of Economics*, 6, pp. 250–78.

Willinger, M., and Zuscovitch, E., 1988, 'Towards the economics of information-intensive production systems: the case of advanced materials', in G. Dosi, C. Freeman, R. Nelson, G. Silveberg and L. Soete (eds) *Technical Change and Economic Theory* London: Frances Pinter.

Wilson, R., 1975, 'Informational economies of scale', *Bell Journal of Economics*, 6, 184–95.

Wiseman, C., 1988, *Strategic Information Systems*, Chicago: Richard Irwin.

Yang, H., Wansley, J. and Lane, W., 1985, 'Stock market recognition of multinationality of a firm and international events', *Journal of Business Finance and Accounting*, 12, (2), 263–74.

Young, S., Hood, N. and Dunlop, S., 1988, 'Global strategies, multinational subsidiary roles and economic impacts in Scotland', *Regional Studies*, pp. 487–97.

Zimmerman, J., 1986, *Once Upon the Future*, London: Pandora Press.

Zis, G., 1984, 'The European monetary system', *Journal of Common Market Studies*, XXIII (1), 45–72.

Index

(NOTE: *Passim* means that the word so annotated is referred to in scattered passages throughout the pages indicated. 'n' means that the entry is contained in a note.)